Piloting at Night

Other Books in the PRACTICAL FLYING SERIES

Piloting at Night

Lewis Bjork

McGraw-Hill

New York San Francisco Washington, D.C. Auckland Bogotá
Caracas Lisbon London Madrid Mexico City Milan
Montreal New Delhi San Juan Singapore
Sydney Tokyo Toronto

Library of Congress Cataloging-in-Publication Data

Bjork, Lewis.
    Piloting at night / Lewis Bjork.
       p. cm.
    Includes index.
    ISBN 0-07-006698-1 (hc).   ISBN 0-07-006697-3 (pbk)
    1. Night flying. I. Title.
TL711.N5B57 1998
629.132'5214—dc21
                                                      98-4236
                                                       CIP

McGraw-Hill

A Division of The McGraw·Hill Companies

1 2 3 4 5 6 7 8 9 0 DOC/DOC 9 0 3 2 1 0 9 8 hc
1 2 3 4 5 6 7 8 9 0 DOC/DOC 9 0 3 2 1 0 9 8 pbk

ISBN 0-07-006697-3 (pbk)
ISBN 0-07-006698-1 (hc)

The sponsoring editor for this book was Shelley Chevalier, the editing supervisor was Caroline Levine, and the production supervisor was Clare Stanley. It was set in the PFS design in Times by Paul Scozzari and Michele Pridmore of McGraw-Hill's Professional Book Group composition unit, in Hightstown, NJ.

Printed and bound by R. R. Donnelley & Sons Company.

McGraw-Hill books are available at special quantity discounts to use as premiums and sales promotions, or for use in corporate training programs. For more information, please write to the Director of Special Sales, McGraw-Hill, 11 West 19th Street, New York, NY 10011. Or contact your local bookstore.

This book is printed on recycled, acid-free paper containing a minimum of 50 percent recycled, de-inked fiber.

In memory of those who were not so lucky . . .

Statistical introduction

A search of 16,841 accidents currently in the NTSB database revealed that 2054 (12%) occurred at night. Of these, 536 were weather related, 583 occurred in the en route phase of flight, and 172 occurred during nighttime instrument approaches. More revealing perhaps is that 214 stemmed from VFR flight into deteriorating weather/IMC, 20 from loss of control during cruise flight, and 76 from VFR descent into terrain/wires/other object(s). (AOPA Air Safety Foundation, Introduction, *Night Flying*)

Contents

CONTENTS

4 Physiology at Night   79

5 Take-off and Climb   105

CONTENTS

6 En route Navigation, Maneuvering, and Weather   119

7 Approach and Landing   141

8 Emergencies   173

Appendix There Are Many Ways to "Go Bump" in the Dark   187

Introduction

This book is written *for pilots* who wish to fly safely at night and therefore takes a pilot's perspective: What is different? What can be done *in the cockpit* to facilitate a night flight? Any related topic that is not specifically aimed at pilots is covered to some extent minimally—is included perhaps only to inform—and our focus instead is the practical knowledge and skills required to successfully operate an airplane at night. With frequent quotes from experienced aviators, accident records, personal histories, federal regulations, and the *Airman's Information Manual* (AIM), each chapter illustrates some crucial methodology for a night pilot to employ in the cockpit. The book is intended as an instructional guide, a practical how-to manual, if you will, for those who are up in the dark.

To facilitate your grasp of the subject, each chapter is broken down into the following sections, where they are appropriate.

STORY

A relevant case study can do more to illustrate the point at hand than almost anything. Reading a case study is similar to gaining experience vicariously from someone who has "been there." All the stories that begin the chapters of this book are based upon fact, a few even have the benefit of explicit firsthand guidance, although I have occasionally taken some dramatic liberties for the sake of retaining your interest. In many cases, the National Transportation Safety Board (NTSB) records provided valuable information in the form of accident synopses, which are included for your edification in the appendix.

BACKGROUND INFORMATION

This section examines the situations that lead up to dangerous events. Understanding why a pilot gets in trouble becomes the basis for developing personal judgment and avoiding similar misfortunes.

INTRODUCTION

TECHNIQUE

This section includes some practices that experienced pilots employ to simplify their flying tasks. They range from something as simple as marking a chart ahead of time or bringing a flashlight to more complicated procedures, such as preparing for a transition to instrument flight rules (IFR) flight when the weather is marginal. The section includes suggestions and advice from numerous airline pilots and flight instructors who make their living by flying in the dark, myself included.

SKILLS TO PRACTICE

Where it is appropriate, a few simple skills in the cockpit can do much to improve your abilities—to make smoother night landings, for example. This section describes a few things that can be readily practiced, enabling you to learn the physical motor skills and gain valuable experience as you prepare for a night flight operation.

FURTHER READING

In researching this book, I was privileged to read a few wonderful books that relate to the subject of night flying. Charles Lindbergh, for example, wrote a beautiful description of his experiences with the Spirit of St. Louis, under that title. There are others. I have included some of these titles for your own enjoyment at the end of each chapter, as they offer more firsthand experiences and detail, if you are interested.

CHAPTER ORGANIZATION

The chapters are organized in a pattern that follows that of a typical flight:

Chapter 1 deals with the go, no-go decision. This chapter can be viewed as a synopsis of the hazards and benefits of a night flight—a list of pros and cons, if you will. Beginning with a frightening situation and ending with a delightful one, it should assist you in weighing the decision of whether to fly in the first place.

The second chapter is a careful study of what may be done, while still on the ground, to prepare for a night flight and offers tips on how to plan the route, what equipment to bring, and selecting appropriate airports.

Chapter 3 is an examination of applicable federal aviation regulations (FARs) as they relate to night flying. Since regulations alone can put me to sleep with the efficiency of a mallet, I have attempted to make a few relevant comments to help you stay awake.

Indeed, staying awake is one of the major physiological difficulties inherent in night flying—it was Charles Lindbergh's greatest challenge. Chapter 4 examines this and other physical factors in some detail and offers several suggestions for training yourself to be more alert and capable in the cockpit.

Having decided to fly, prepared yourself, and understood the legalities and physical limitations involved, the next step is to climb into the airplane and take off. Chapter 5 reveals some of the dark hazards lurking just off the departure end of the runway.

En route hazards and weather are the basis for Chap. 6. Sometimes it is best to go "IFR," that is, "I follow roads," where a night cross-country is involved, or, perhaps, you might actually fly IFR, as determined by your judgment of the weather.

Night landings are a story in themselves, causing perhaps the greatest fear and misfortune among pilots. The night approach and landing is the subject of Chap. 7.

Chapter 8, after all this careful preparation, looks at what you might do when something goes wrong. After all, *if anything* can *go wrong* . . .

As with every endeavor as potentially serious as flying, this book alone cannot fully prepare you with the required skills to safely tackle a night flight. You have to practice. If you are uncomfortable with some of the concepts described herein, you might benefit from the guidance of a capable flight instructor and a beautiful evening. I wish you many happy returns in your efforts.

Happy flying.

Lewis Bjork

Acknowledgments

Thanks to my wife, who edited this book and accompanied me on photographic journeys into the darkest of nights; she managed home and children largely without my help until this project was completed.

Thanks to my family and extended family for their encouragement; to Paul Smith, again, for preventing my computer from swallowing its files, and a lovely evening dinner at Wendover; to fellow pilots at Skywest who contributed dramatic material, and made helpful suggestions as we developed the list of pros and cons; to Richard Pratt for donating the use of his Mooney for some night photography; and finally to all those pilots whose unfortunate experiences shed valuable light on the subject of night flight.

Abbreviations

3-D	three-Dimensional
ADF	Automatic Direction Finder
AFD(s)	Airport Facility Directory(s)
AIM	Airman's Information Manual
AIRMET	in-flight weather advisory
ALS	Approach Lighting System
AOPA	Aircraft Owner's and Pilot's Association
ARTCC	Air-Route Traffic Control Center
ATC	Air Traffic Control
ATP	Airline Transport Pilot
AZ	Arizona
CAVU	Ceiling and Visibility Unlimited
CDT	Central Daylight Time
COM	COMmunications
CST	Central Standard Time
DME	Distance Measuring Equipment
EDT	Eastern Daylight Time
ELT	Emergency Locator Transmitter
FAA	Federal Aviation Administration
FAR(s)	Federal Aviation Regulation(s)
FBO(s)	Fixed Base Operation(s)
FMS	Flight Management System
FSS	Flight Service Station
ft	feet
GPS	Global Positioning System
Hg	Mercury (the metallic element)
hr	hour

ABBREVIATIONS

HST	Hawaii Standard Time
IFR	Instrument Flight Rules
ILS	Instrument Landing System
INS	Inertial Navigation System
IRS	Inertial Reference System
kts	nautical miles/hour
LORAN	LOng RAnge Navigation
MDT	Mountain Daylight Time
MSL	Mean Sea Level
NASA	National Aeronautics and Space Administration
NAV	NAVigational
NAVAIDS	NAVigational AIDS
NOAA	National Oceanic and Atmospheric Administration
NOS	National Oceanic Service
NTSB	National Transportation Safety Board
PAPI	Precision Approach Path Indicator
PAR	Precision Approach Radar
PDT	Pacific Daylight Time
REIL	Runway End Identifier Lights
RJ	Canadair Regional Jet CL-65
RNAV/ARNAV	ARea NAVigation
SIGMET	Inflight weather advisory of serious nature
U.S.	Corporate United States
UFO	Unidentified Flying Object
UT	Utah
VASI	Visual Approach Slope Indicator
VCR	Video Cassette Recorder
VOR	Very high frequency Omnirange Receiver
WAC	World Aeronautical Chart
X	distance in feet

Chapter 1
Up in the night

Max and Jim flew a rented Cessna 172 around spectacular Hawaiian scenery at sunset. Their wives were in the back. Gorgeous vistas and a spectacularly rugged coastline stunned the occupants of the little plane, putting everyone in a state of rapturous awe until it got dark—sunsets do that. They flew over ocean water on a moonless night, single engine and happy under stars that shone with the true brilliance of a billion distant suns, undimmed by city smog. Despite the noise of the Cessna's engine, the cockpit, somehow isolated from the world outside, rumbled away peacefully, almost uplifting—duplicating the stunning scenery slowly fading to night outside. The starlight reflected from the ocean waves, twinkling from above and below with a horizon that stretched into infinity.

As the light fades, so does the ocean horizon (Fig. 1-1). Its color is a perfect match for the night sky, and without the glimmer of moonlight, the horizon is impossible to see. The lights of cities on the islands soon appeared to float in space without perspective. For all the pilots knew, they could be upside down, or standing on end a few feet away—there simply was not any useful visual reference. In the comforting, inward solitude of the aircraft cabin, the panel glowed softly red, illuminating the artificial horizon, turn gyros, and other primary gauges. Max and Jim were both instrument rated, and the plane certified for flight in the clouds. They had no worries, no feelings

Fig. 1-1. *An unmistakable line in the day, the horizon disappears at night.*

of trepidation. They were completely comfortable as they headed back around the island, navigating by starlight and glittering shoreline.

The airport had a few low clouds and they planned to use the instrument approach to end a perfect flight on a perfect evening. As they continued, conversing quietly over the intercom, the lights outside appeared fuzzy and began blinking out. With no fanfare and no warning, what was left of the island view simply disappeared, as the little plane entered a layer of stratus clouds. The air remained smooth, but felt a bit more humid. Other than that, there was no evidence of anything beyond the cockpit itself, which was completely enveloped in blackness. With nothing to see, and their workload increased as they transitioned completely to instruments, conversation in the cabin faded away to silence.

A bright red light flashed on the panel about then, indicating that the alternator had failed. In daylight, this would present a benign problem; but at night, it marked the beginning of a silent countdown that could end with a fatal crash. The alternator provides electrical energy for lighting, radios, navigation instruments, and a few odd busses and switches. It is backed up by the battery, which can maintain electrical power to the system for an unknown time, ranging from a few minutes to about an hour—depending on the quality and charge of the battery. The system, and everything it powers, would be worthless when the battery drains out. The fact that the plane flew in IFR conditions placed a priority on a few components of the electrical system—namely, the turn and bank gyro, the radios, and the cockpit lights. Thankfully, the other gyros are not powered by electricity, but by suction—as long as the prop turns outside, at least *those* gyros should work.

They attempted to reset the alternator switch, but it refused to come on line. On battery power only, they might be without an electrical system before they could use the instrument approach equipment at the airport, still more than 30 minutes away. The plane would be difficult enough to fly enroute, with the electrics shut down—dead reckoning with a partial panel, across water and mountains to a small island in the middle of thousands of square miles of ocean—but they felt it could be done, especially if they could power up the navigational (NAV) radio once in a while and take a position fix. They discussed the situation, being careful not to alarm their wives.

"It's only an alternator light. We need to conserve the battery power, so we'll turn off all the unnecessary systems…there go the external lights…the COM radios, the standby NAV radio, the DME, the ADF, and—sorry—the aft cabin lights," Max said, as the cabin darkened, with only a dull glow coming from the instruments. They stopped at one NAV radio and the master switch. He looked at Jim—"Ready?"

"Might as well." Max flipped the master switch off and the cabin went black. Pitch black. For a moment the occupants of the plane were stunned. Like darkness within the depths of a subterranean cave, they couldn't see a moose in front of their faces—they couldn't see squat; and, to their dismay, nobody had thought to bring a flashlight. With a simple decision to turn the master off and conserve power, they overlooked the obvious fact that it was dark—really dark. Without electrical power, the instrument lights go out. Completely blinded by the night, Max fumbled for the master, and the lights came back on. He took a moment to check the plane's attitude and altitude—and correct the bank.

Everyone searched the cabin in vain for a flashlight. They would have to shut the electrical system down and conserve its power, or an instrument approach would be impossible without radio navigation equipment, or at least a radio and some assistance from a controller. The system would keep if it were powered down until needed for the approach, but left on, well, no light-plane battery would last that long. They would be able to fly around for a little while, until the battery went dead and the lights faded away; then, instruments becoming invisible in total darkness, they would die, perhaps spiraling out of clouds into the dark sea, or crashing in flames against one of the island cliffs. With this realization, the mood in the little airplane suddenly turned cold and tense. If they could not find a way to see in the dark within the next few minutes, they could be dead.

A light—any light…got a light? Then Max remembered his butane cigarette lighter. He fished it out of his pocket and flicked it on. It immediately blew out in the breeze coming from the cabin vents. Jim closed the vents. Max snapped it on again. The little flame appeared feeble and fragile in light of their predicament—hardly more than a candle. As Max held the lighter, Jim reached across the center console for the master switch and smiled a little. It seemed funny to be flying IFR by the light of a Bic—it'd make a good commercial.

"Ready?"

Max nodded, and Jim shut the system down. Suddenly the cabin appeared to flicker and dance, reflecting the pilot's faces as if from a campfire. Without that tiny flame, they would die. Jim flew the plane while Max held the lighter and kept it burning. Suddenly, it went dark again. It was as if the engine had quit; the only audible sound was a scratching made by the lighter as Max attempted to restart it. An instant later, they could see again. Jim adjusted the attitude and checked—the engine was running fine.

"Burned my thumb...sorry," Max said.

They found themselves leaning forward close to the panel, sweating in the stagnant cabin air and frequently checking the clock, blindly hoping to avoid mountains and sea alike. They flew like this for 20 minutes, with their wives sitting like silent ghosts in the back seat. They hoped that enough time had passed to bring them within range of the airport VOR. Max flipped on the master switch and doused the lighter, sucking on his burned thumb. The VOR gave a strong signal, so they flipped on the distance measuring equipment (DME). Pretty close to the initial approach fix, less than 15 miles away. They flipped on the COM radio and asked for a weather advisory; then they announced their intentions to make the approach, hoping that the battery would hold out. Ten minutes later, they intercepted the inbound course and throttled back to begin a descent. The instrument lights were noticeably dimmer. Keying the microphone a few times, they hoped the radio had sufficient power remaining to activate the runway lights. This realization came with a start. What if they broke out of the clouds and couldn't see runway lights? They continued the descent with mounting anxiety; then, as if drawing back a black curtain, lights appeared below. The lights of a runway.

After landing, Max could have kissed his lighter but didn't—it was still hot.

BACKGROUND INFORMATION

Max and Jim had a close call in the night, which would have been relatively benign in daylight. Darkness, however, made the situation potentially fatal (Fig. 1-2). Without light, they were almost sure to get vertigo en route and crash, unable to determine which way is up, unable to see their attitude. For a few moments, they were frantic for light. To see is to live. Fortunately, this once, Max was a smoker—he has since quit but now carries a flashlight.

Why Go Up in the Night?

Bats, owls, and stealth fighters prefer to fly at night, taking easy advantage of surprised prey. The owl has specially adapted eyes and quiet wings, the bats use some highly developed radar, and the stealth pilot has the advantage of very expensive night specific machinery. In these three cases, daytime flight is eschewed for the more serious advantages of a night attack.

More common night flying operations include night freight pilots who have literally not seen the sun from the windows of an airplane for years. There are also many cities located at the extreme latitudes where airplanes are a critical part of the infrastructure; daytime operations are impossible for the entire winter because they fly where the sun doesn't shine.

The most compelling reasons for night flight are usually financial, and they are weighed carefully against some of the serious safety concerns. The following section introduces some of the pros and cons of flying in the dark, and each of the points raised here will be discussed with more detail in later chapters, as indicated. You may find that although the night flying question raises serious safety issues, it also provides very real advantages. In the end, the choice to fly in the dark is yours.

Fig. 1-2. *Commercial helicopter tour operations at this island airport are illegal after dark.*

Increased Aircraft Utility

It is impossible to ignore the economic benefits of night flying. From the professional standpoint, airplanes make money only when they fly, and the expenses of ownership are reduced by keeping the plane in the air as often as possible. Avoiding night operations cuts the useful benefits of owning an airplane in half. However, flying the airplane both day and night maximizes financial returns because the machine is in operation a greater percentage of the time (Fig. 1-3). Few competitive aircraft operations can afford to ignore the need to fly and make money in the dark. The only exceptions to this rule are flight operations for which the risk of night flying exceeds the benefit; crashed machinery costs more money than parking the plane overnight. An example of high-risk flying is bush operators, who operate primarily out of unlighted runways. When the sun goes down, using those runways becomes hazardous. The hazard is enough to make the financial advantages of night flying pale in comparison to the potential risk of a crash.

Nevertheless, nighttime is often a business's best opportunity to get ahead of the competition. The mail system, for example, works well into the night to facilitate a daytime delivery. Many freight companies have successfully capitalized on the concept of overnight delivery, employing airplanes and flight crews almost exclusively in night operations to complete their deliveries by the start of normal business hours the next day. Businesses pay a premium to these companies for the time advantages provided by their overnight service. Cropdusters venture into the dark with the aid of a powerful system of

Fig. 1-3. *The most compelling reasons for flying at night are usually economic.*

lights. The busy dusting season is simply too short to cover all the acreage that farmers demand unless the pilots can fly well into the night.

For the general aviation pilot, night flight has similar advantages. Consider the aerial commuter. The pilot flies to work in the morning, grateful for the time-savings provided by the airplane. After a full day's work, the pilot goes back to the airplane and flies home—and probably does that in the dark; capitalizing on a day in the office means that the home commute waits until business is complete. Why cut the business day short in order to fly home in daylight? During winter, days become shorter still. Avoiding night flying in the colder months would require a late arrival to and a yet-earlier departure from work, meaning a *shorter* business day, thus negating the advantages of flying to work in the first place. Furthermore, if the pilot combines a need for daylight hours with an equally strong need for good weather and ideal operating conditions, the airplane's use is reduced to the level of a hobby or sport, negating much of its transportational value.

The Air Is Often Smoother at Night (Chap. 6)

The weather is driven by differential heating of the atmosphere. The sun heats the equatorial regions more than the poles and thus causes the atmosphere to circulate. On a smaller scale, parking lots get a little hotter than pine forests, and hot air rising above industrial areas can create rather powerful thermals and other "convective" activity. One well-known example of solar-powered weather is a thunderstorm.

When the sun goes down, the air cools, and convective-type weather slows or may even die away completely. This works to the pilot's advantage when flying at night. Late-afternoon thunderstorms and their associated turbulence are often long gone by the darkened hours of early morning; severe clear summer skies turn progressively to cumulus buildups, followed by scattered thunderstorms, and then nighttime. A pilot flying throughout the day will experience a progressively bumpier ride as the convective systems develop. When night falls, the clouds begin to cool and lose energy. The once rising air settles back down, and the sky is clear again by morning. The pilot flying into the night will experience a progressively smoother ride, which fades to glass by sunrise (Fig. 1-4).

Convective activity aside, the lack of sunshine will not make all weather hazards go away. Many kinds of clouds and weather conditions are very stable and don't require sunlight to fuel them. Wind is a good example of this—it is most often pressure related on a large scale and may blow steadily for days in a prevailing direction. Stratus-type clouds and large-scale weather patterns are affected much less by sunshine at the local level and may exist for days at a time. However, the hazards presented to a pilot by stable weather conditions are somewhat less threatening than thunderstorms, and real dangers like icing in clouds tend to be more localized at night. The air is simply not moving as much to mix things up, so the night pilot has another advantage here.

Better Takeoff and Climb Performance (Chap. 5)

One basic rule of weather is that when the sun goes down, the air cools. Cool air is more dense than warm air and makes better compression in airplane engines and better lift

Fig. 1-4. *Weather systems are solar powered. They lose energy after sunset.*

across the wings. A plane may be expected to take off shorter and climb better than it would in the daytime, sometimes by a significant factor.

Lighted Objects Visible at Great Distances (Chap. 4)

Your eyeball is tuned to see reflected light. Light bouncing off this page is collected and focused upon your retina, and you read. Turn the light off, however, and you don't read much at all, let alone see the book. If the page were lit internally, say, like neon lights, you could read it from some distance in the dark. In that case, the page is emitting its own light and becomes highly visible. Interestingly, a neon sign in daylight pales in comparison to sunshine and looks comparatively dull, like a star—or an airplane with a full complement of recognition lights. You may hardly see the airplane with its lights on in the daytime—it is a small, rather dimly lit object flying a few miles away. Put the sun out, however, and the plane becomes a glowing beacon against a dark background, visible at much greater distances. The night pilot may see other aerial traffic and remote destinations at outrageous distances—traffic and places that would be totally invisible in daylight (Fig. 1-5).

For the high-flying pilot on a clear night, cities, roads, and other lighted landmarks become far more readily apparent. Normal daytime glare fades when the sun goes down, and visibilities far in excess of a hundred miles are possible in clear air. Astronauts orbiting the earth, for example, can see pitifully little of humanmade creations from their daylight vantage point. It takes something as vast as the Great Wall of China to show up in the naked eye from over 100 miles high. When the astronauts fly to the dark side of the earth, however, cities become a sparkling network of lights far more beautiful than they appear from close range. So, when a plane or object is lit, the night pilot can see it from much greater distance than a pilot in the daytime.

Fewer Bird Strikes

Most birds avoid flying at night and those that do fly are rather good at it, so I suppose that the chances of a bird strike are less in the dark.

Less Congested, More Convenient Air Traffic Control

Nighttime air traffic is similar to nighttime auto traffic on a highway: For the most part, there is less of it. Keeping that in mind, there are fewer highway patrol officers out there, too, because the majority of people sleep at night and do their traveling during the day hours. For the night pilot, the skies become less crowded. There is less chatter on the radio, and air traffic controllers are often more relaxed. Several control towers close down for the night, because there simply is not enough traffic to justify the need for a controller to be on duty. Victor airways and jet routes open wide to the IFR pilot, making direct clearances much easier to obtain and providing fewer traffic-related delays.

Some places never sleep, however. Just as highway traffic remains congested or even dangerous at all hours in big cities, air traffic to those airports does so as well. A pilot waiting in line to take off from Los Angeles airport on a (rare) clear night will often see airplanes on approach strung out like Christmas lights for 40 miles or more. Air traffic controllers at places like this are often as busy in the night as they are in daylight, except it is harder for them to stay awake. Nevertheless, these big cities are thankfully still the

Fig. 1-5. *I-80 Eastbound as seen from a jet cockpit at about 18,000 ft. The city on the "horizon" is over 80 miles away—the stars are a little further.*

exception to the rule—most of the world slows down and sleeps when the sun does, thus offering a convenience to the dedicated night flyer.

Faster Service at 24-hr Fixed Base Operations
One convenience that stems from the less congested night environment is faster service at all-night fixed base operations (FBOs). A corporate pilot looking for a quick-turn at an intermediate fuel stop may have much better luck in the nighttime because she or he may be the only one there. A general aviation pilot in the same situation should have better access to courtesy cars, quicker refueling turnarounds, and generally better service.

Easier Access to Rental Planes
At one flight school where I worked years ago, the night service was do-it-yourself. A pilot wishing to rent a plane had only to walk in, grab the keys, and drop the rental fee into a slot after landing. There was no finagling for available planes, and little or no

advance scheduling was required. Compared to the usual daytime congestion and hassle, that's as convenient as it gets. On the downside, some flight schools require written permission for any night rentals or operations, killing the convenience of spontaneity altogether; nevertheless, the possibility of an easy night rental is out there.

Great Excuse to Wear the Leather Jacket

Since nighttime is generally cooler than daytime, it's an opportunity to wear your favorite flight jacket—although the matching Ray-Bans will render a pilot practically blind. You can still hang the shades over your breast pocket, however, and your glow-in-the-dark pilot watch will look really cool.

Moonlight and City Lights Make for Fantastic Scenery

City lights from a dark aerial vantage point can be positively romantic, and passengers often become speechless when dazzled by a gentle night flight over a populated area. Cities that are gray and industrial in the daylight become scattered stars and fireworks, jewels over black velvet, by night. The lights flicker, flash color, and move, dancing with change and depth below the airplane, reflecting off wings and clouds. The aerial transformation of a city at night can be magical and stunningly beautiful (Fig. 1-6).

However...

The major difficulty surrounding night operations is the fact that the human eye does not see particularly well in the dark. Our eyesight depends on the existence of reflected light

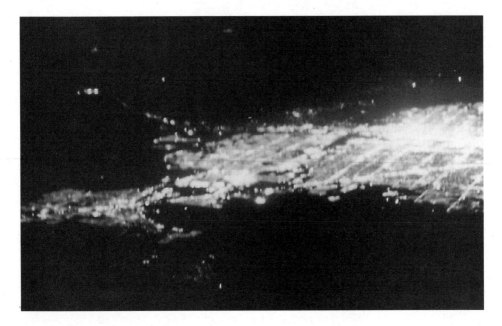

Fig. 1-6. *Las Vegas, Nevada, from 35,000 ft.*

from the sun, moon, or artificial sources, without which we're as blind as bats with broken radar. This is a serious problem in an aviation environment. A pilot's eyes are critical to his or her ability to fly the plane. For example, try sitting in a cockpit with your eyes closed and use a sectional chart to find the latt/long coordinates for the tallest obstacle within 5 miles of your home airport. This is an impossible task, unless you're peeking or very lucky. Now, to make it simpler, keep your eyes shut and just point in the direction of the tallest obstacle near your airport (as you don't see it out the windows)—also an impossible task to do accurately.

The bottom line is this, you have to see to be able to fly the plane, and it is more difficult to see after the sun goes down. Almost all night-flying hazards stem from this one major problem.

Unlighted Objects Are Invisible (Chap. 4)

In darkness, the human eye is blind to anything that does not emit its own light. Therefore, to the pilot, the unlit world is useless. Common objects unmentioned and no problem in daylight become far more hazardous and frightening at night. Unlit airplanes are invisible, a fact used to the advantage of smugglers attempting to breach guarded borders. Since the pilot rarely *hears* other planes in flight, there would be no warning of a potential midair collision with an unlighted airplane.

Large bodies of water become black in the night, occasionally reflecting moon or starlight. If the water reaches or perhaps makes the horizon, like the ocean, darkness will cause the horizon to disappear, making VFR flight by outside references a practical impossibility.

For pilots used to navigating mountain passes in daylight, darkness causes the mountains to become invisible barriers of rock, impossible to navigate at close range and utterly dangerous in an emergency.

Bad weather that is perhaps easily circumnavigated in daylight also becomes invisible at night. The unsuspecting pilot may realize the airplane has penetrated clouds only by the fact that the city lights below suddenly disappear. There is often no other indication until the airplane actually encounters the storm.

Although many wires and powerlines that exist in close proximity to runways carry some kind of visual aid, perhaps even lights, the vast majority are unmarked and from the air are difficult to see even in the daylight, let alone at night. These become a factor in flight at low altitudes, and such flight becomes extremely hazardous at night without proper equipment. Trees, except for Christmas time, rarely carry lights of their own and become another nighttime obstacle that is impossible to see and avoid at low altitude. There are innumerable other possibilities for hazard in darkness: Suppose a pilot knows the area well enough to be sure of the location of a particular field and, landing there, clobbers an invisible cow. Runway obstacles that are easily avoided in daylight require glidepath lighting or an instrument approach at night. And, of course, there are antennas, towers, dark buildings…this list could go on forever.

Visual Illusions (Chaps. 4 and 7)

Since your eyes are severely disadvantaged in darkness, they're easily fooled. You see a single beacon in the darkness, is it a star, an airplane, or something else? The light appears to move, so it must be another airplane. Is it really? Or is it a star that "appears" to move because your own plane is turning? Since your brain references a great deal of subtle visual details in order to stabilize and add depth to what you see, a little turbulence and dim lighting may cause your visual picture to become blurry (Fig. 1-7).

Your depth perception depends greatly on the clarity of the image you see. This clarity diminishes in darkness, and you lose a lot of accuracy in judging distance. Mistaking a star for another airplane, for example. The star is billions of miles away, the plane would be relatively close—and you might not be able to see the difference.

Vertigo (Chaps. 4 and 5)

The human body has a simple balance indicator. The fluid sitting in canals in your inner ear can sense acceleration forces, and these sensations are cross-referenced with what you see. You usually have a good idea of where "down" is because of the gravitational influence on the fluid in these canals. That is why you don't fall over when you shut your

Fig. 1-7. *Flight instruments are more difficult to read at night in turbulent air.*

eyes. If your internal "gyros" are tumbled because you've been riding the Colossal Super Looper five times straight, your eyes can usually help you stay upright, even though you might stagger a little because of the scrambled fluid in your ears. Eyes and ears work together to keep you upright. In many people, there is a connection between the eyes, ears, and the stomach too. If the first two disagree, the stomach throws in the towel by vomiting at the first sign of conflict.

An airplane has the ability to generate acceleration forces in any direction. There is little to stabilize your personal gyros in the absence of visual indications, as in darkness. The natural result is a complete inability to discern which way is "up" and a very real possibility of motion sickness.

Inadequate Visibility for an Emergency Landing (Chap. 2)
Unlit objects are invisible without ambient light. These objects include fields, houses, horses, trees, power lines, and mountains—to name a few. The chances of finding an ideal place to conduct an emergency landing in the darkness become reduced to the equivalent of dumb luck. There is a famous aviation saying that goes something like this: "If you lose your single engine at night, you should set up a controlled glide. When you think you are close to the ground, turn the landing light on. If you don't like what you see, turn it off." Many pilots simply refuse to pilot a single-engine airplane at night without a parachute because the chances of making a safe emergency landing are so poor (Fig. 1-8).

An obvious solution to the question of night emergency landings lies in a second engine. Theoretically, the occupants of a multiengine airplane that can continue flight after an engine failure will never have to face an emergency landing on an unlit runway. Theoretically.

Odds Are against You
All this difficulty with vision at night results in some very real numbers. According to NTSB records for the last 11 years, you are a little more than twice as likely to die if you crash your plane at night as opposed to a similar crash in daylight. For a breakdown of the NTSB figures, see Appendix A.

Automatic Rough—Higher Pilot Anxiety (Chap. 4)
If you find yourself flying at night, or in any other condition where an emergency landing would be unsafe (over water, rugged terrain, etc.), your anxiety levels noticeably increase. You listen to the engines more carefully than usual and, because of this, hear sounds you hadn't noticed before. Is it normal for the engine to be knocking like that? Where did that little chatter come from? This is "automatic rough."

Most pilots consciously survey the engine conditions before committing to fly over forbidding terrain, and do it almost constantly at night. While doing so, their levels of stress increase—they experience higher anxiety and are generally less comfortable with their situation.

Smaller Airports Closed, No Gas or Services
Although an airport might have pilot-controlled runway lighting, there might be nobody there to greet you when you land. People generally go to sleep at night. Public services slow down. Fuel pumps might be off, restaurants closed, and the bathrooms locked. Hopefully the pay phone is outside. Calling the airport manager out at this hour might

Fig. 1-8. *Landing lights have a dismally short reach.*

impose a hefty callout charge on top of the fuel bill. This is a typical situation for most of the lighted airports in the world.

Many airports impose a noise curfew for operations at night, limiting operations almost completely during the hours that people want to sleep.

For the traveling pilot who is in need of services and a rental car, the night destinations could easily be restricted to rather large airports with all-night services, consigning the pilot to areas with more noise and higher traffic.

Unlighted Runways Not Safe (Chap. 7)

Airports that have no source of runway lighting become quite dangerous in darkness. This includes the majority of grass runways and mountain airstrips—almost every "bush" airport. If you're camping from your airplane in the wilds of northern Idaho, for example, nighttime flight in the region can turn the most picturesque and inviting runway into a black hole full of hazard and insecurity, with crushing, invisible mountain rocks lurking in the shadows.

Fatigue (Chap. 4)

There is a rhythm to the human sleep-wake cycle, and it's pretty much what you'd guess—about 8 hours of sleep in every 24. If the need to sleep is ignored, the body can be temporarily ruined. Mental acuity decreases, coordination falls apart, judgment is impaired, and a strong desire to sleep is felt. If a tired pilot ignores these indications for, say, another day or so, the body will sleep anyway. The pilot may experience hallucinations—dreaming with eyes open—and complete loss of control and useful consciousness. Sleep is *required.*

A pilot flying at night on a regular basis can adapt these sleep-wake cycles to being alert during dark hours. A pilot used to sleeping at night, however, may experience a strong need for sleep and its consequences when flying at night on irregular occasions.

Hazardous IFR Conditions in Visual Weather (Chap. 6)

On a moonless night over sparsely populated terrain, a pilot might have severe difficulty finding a visual reference to use as a basis for controlling the aircraft attitude—encountering IFR conditions in million-mile visibility. The horizon can become invisible, and lights below in many areas of the world are sparse enough to be useless or even dangerously misleading. Stars can often get confused with sparse ground lighting. Large areas devoid of ground lights include large bodies of water and mountainous terrain, deserts, tundra, and jungles—basically, anywhere that the human presence is scarce (Fig. 1-9).

Flight over these regions would be impossible on a dark night unless the airplane had at least basic gyroscopic instruments and a pilot skilled in flying by reference to instruments. The use of pilotage for navigation would become impractical in these circum-

Fig. 1-9. *Large uninhabited areas may produce dark IFR conditions.*

stances, creating a need for dead reckoning or some sort of radio, satellite, or celestial navigation reference, with corresponding equipment in the airplane, as well.

Much Better Weather Required for VFR

Since clouds and weather are unlit (except for lightning in thunderstorms) and often invisible in darkness, the option of scud-running after sunset becomes useless. Furthermore, since the pilot is unable to avoid hazards that are invisible, and thus easily maintain VFR conditions at night, the acceptable weather minimums for a VFR flight drastically increase (Fig. 1-10). If the pilot is restricted to VFR because of personal or equipment limitations, almost *any* weather forecast along the route of flight could be cause for a "no-go."

Special VFR Impossible without IFR Capability

Special VFR is basically a scud-running approach technique. A pilot attempting to scud-run after dark is just begging for inadvertent IFR conditions. Consequently, special VFR at night (1-mile visibility and clear of clouds) is legally restricted to IFR-capable combinations of plane and pilot (Fig. 1-11).

No Tower and Fewer Emergency Services

Many control towers are closed during nighttime hours. Air traffic is generally much less during these hours, reducing the need for a tower controller, but the pilot loses the benefits of having an extra set of eyes on the ground. Also, the emergency equipment (firefighting vehicles and crash trucks) might not be crewed at smaller airports at night for the same reasons. If an airplane crashes after hours, it might be a few minutes be-

Fig. 1-10. *Weather may be invisible at night.*

Fig. 1-11. *On final for Salt Lake City in marginal VFR conditions. The break in the clouds ahead is made by the wake disturbance of preceding aircraft. The same conditions at night would require instrument equipment and a rated pilot.*

fore the local fire department can arrive on the scene—perhaps even quite a while, because the crash might go unreported with nobody at the airport.

Electrical Failures Far More Serious

One of the key functions of an electrical system is to provide lighting. If the system fails—alternator, battery, short, or fire—the lights go out, and the pilot could potentially "go blind." This is not a big deal in daylight, since the pilot can see to fly the plane. At night, however, the pilot might be unable to see adequately to do anything. Instrument flight could become impossible. This is a grave emergency. A working flashlight becomes an invaluable tool in this case—something not to be forgotten (Fig. 1-12).

Depth Perception Difficult to Judge

Your eyes use a lot of little visual cues to judge depth and distance. You unconsciously compare the relative sizes of objects in the distance and gauge their motions and subtle differences in appearance. These images are put together in your brain in a way that you get a nice three-dimensional picture that is helpful in landing your airplane. Nighttime removes most of these visual cues. You still see lights, but distance becomes relatively difficult to accurately judge. You may have a nasty time flaring properly or landing the plane altogether. Other air traffic could be much closer or further away than you think. The whole visual world is potentially deceptive at night.

Fig. 1-12. *This could make the difference between success and total failure.*

You'll have difficulty gauging the progress of a moving target outside the plane. At night, the motion and range of air traffic that is moving directly toward or away from you is far more difficult to detect, and traffic below, against a background of city lights, may be invisible, even at very close range.

Trickier to Land—High Performance a Challenge

Night landings are relatively difficult. If the runway is short and approaches the minimum required for the airplane, the pilot has a very difficult task. Simply touching down in the first useful portion of asphalt might be impossible, because the pilot has to adhere to visual approach slope indicator (VASI) or other glide-slope indications to avoid the "invisible" hazards during the approach. This night factor could significantly lengthen the runway required for your airplane beyond what is specified in the performance charts.

Soft-field landings which require a deft touch of the pilot may become very difficult because the pilot's perception of depth is reduced, requiring landing techniques which increase the required runway length.

More Equipment Required

Night flight legally requires lights and, in many cases, several instruments that are not required in daylight. For initially outfitting the plane, night capability is more expensive.

More Demands on Pilot

This goes without saying.

Search after Crash Requires Daylight

You are probably an "unlit" object after an emergency landing. Search and rescue teams need daylight to find you after a crash. Searching at night would be futile. If a night crash has left you with a dire need for help or medical attention, you have to wait for morning before the search for you can really begin—even if they know you've gone down.

More Reserve Fuel Required

The reserve fuel you must legally carry at night increases by 50 percent over daytime reserves. This limits your payload and may reduce the number of passengers you could legally carry.

Cooler Air May Be Too Cold

We talked of cold air as a benefit to takeoff and climb performance, but cold air can be a hazard to your personal comfort. If you crash in the desert, for example, nighttime brings the possibility of freezing to death—even during the warm seasons.

Carburetor Ice

Temperature and dewpoint spreads narrow in the nighttime cool, saturating the air and giving rise to ground fog and low visibility hazards as well as a proclivity to conditions ideal for the formation of carburetor/induction icing.

TECHNIQUE

At this point, the choice of night flight is yours. There are probably more pros and cons that could be discussed, but you have just read the meat of the question. Assuming you have made the decision to join the rest of us who are up in the night, the following chapters will concentrate on what you may do differently in the cockpit to stack the pros in your favor, thereby avoiding the nastier cons.

From "Night Flight"

He glanced back toward San Julian; all he now could see was a cluster of lights, then stars, then twinkling star-dust that vanished, tempting him for the last time.
"I can't see the dials; I'll light up."

Chapter One

He touched the switches, but the red light falling from the cockpit lamps upon the dial-hands was so diluted with the blue evening glow that they did not catch its color. When he passed his fingers close before a bulb, they were hardly tinged at all.

"Too soon."

But night was rising like a tawny smoke and already the valleys were brimming over with it. No longer were they distinguishable from the plains. The villages were lighting up, constellations that greeted each other across the dusk. And, at a touch of his finger, his flying-lights flashed back a greeting to them. The earth grew spangled with light-signals as each house lit its star, searching the vastness of the night as a lighthouse sweeps the sea. Now every place that sheltered human life was sparkling. And it rejoiced him to enter into this one night with a measured slowness, as into an anchorage.

He bent down into the cockpit; the luminous dial-hands were beginning to show up. The pilot read their figures one by one; all was going well. He felt at ease up here, snugly ensconced. He passed his fingers along a steel rib and felt the stream of life that flowed in it; the metal did not vibrate, yet it was alive. The engine's five-hundred horse-power bred in its texture a very gentle current, fraying its ice-cold rind into a velvety bloom. Once again the pilot in full flight experienced neither giddiness nor any thrill; only the mystery of metal turned to living flesh.

So he had found his world again....A few digs of his elbow and he was quite at home. He tapped the dashboard, touched the contacts one by one, shifting his limbs a little, and, settling himself more solidly, felt for the best position whence to gage the faintest lurch of his five tons of metal, jostled by the heaving darkness. Groping with his fingers, he plugged in his emergency-lamp, let go of it, felt for it again, made sure it held; then lightly touched each switch, to be certain of finding it later, training his hands to function in a blind man's world. Now that his hands had learnt their role by heart, he ventured to turn on a lamp, making the cockpit bright with polished fittings and then, as on a submarine about to dive, watched his passage into night upon the dials only. Nothing shook or rattled, neither gyroscope nor altimeter flickered in the least, the engine was running smoothly; so now he relaxed his limbs a little, let his neck sink back into the leather padding and fell into the deeply meditative mood of flight, mellow with inexplicable hopes.

Now, a watchman from the heart of night, he learnt how night betrays man's presence, his voices, lights, and his unrest. That star down there in the shadows, alone; a lonely house. Yonder a fading star; that house is closing in upon its love....Or on its lassitude. A house that has ceased to flash its signal to the world. Gathered round their lamp-lit table, those peasants do not know the measure of their hopes; they do not guess that their desire carries so far, out into the vastness of the night that hems them in. But Fabien has met it on his path, when, coming from a thousand miles away, he feels the heavy ground-swell raise his panting plane and let it sink, when he has crossed a dozen storms like lands at war, between them neutral tracts of moonlight, to reach at last those lights, one following the other—and knows himself a conqueror. They think, these peasants, that their lamp shines only for that little table; but, from fifty miles away, someone has felt the summons of their light, as though it were a desperate signal from some lonely island, flashed by shipwrecked men toward the sea. (Antoine de Saint-Exupéry, *Airman's Odyssey,* Harcourt Brace Jovanovich, Orlando, FL, p. 215.)

SKILLS TO PRACTICE

It all starts with the planning....

FURTHER READING

de Saint-Exupéry, Antoine, "Night Flight," *Airman's Odyssey,* Harcourt Brace Jovanovich, Orlando, FL, p. 215.

Chapter 2
Planning

And all those myriad lights, all the turmoil and works of men, seem to hang so precariously on the great sphere hurtling through the heavens, a phosphorescent moss on its surface, vulnerable to the brush of a hand.
I feel aloof and unattached, in the solitude of space.

Charles Lindbergh, *The Spirit of St. Louis,* **Charles Scribner's Sons, New York, 1953, p. 11.**

The pilot has long settled into the wicker chair of his cockpit. Not exactly cramped, the cabin dimensions exactly fit his person with enough room to work and no room to spare. San Diego is more than 4 hours behind, St. Louis, about 10 hours further on. He'll be comfortable for the duration, staring at a sparse panel, listening to the rumble of the plane's single engine, and watching the world roll by outside. It's 7:52 p.m., over the deserts of Arizona; the last vestiges of twilight fade away to the rear as the plane motors eastward into a sky the color of deep sea, plunging through the mouth of night's dark tunnel towards the dawn—almost 9 hours away. Details on the ground outside fade with the dwindling light until only the barest ghosts of a huge desert remain. The pilot is thoughtful of his circumstances, considerate of his ambitions, and contemplating the vastness of the country, when the engine skips a beat. Like a conductor, hearing a sour note, or miss-timed refrain, the pilot cocks his attention, suddenly uncomfortable at the discordant sound. The engine coughs again…and begins to shake.

The beam of his flashlight reveals adequate fuel pressure, normal parameters on the other gauges—nothing obviously wrong, but the sick engine is shaking the entire plane. Without sufficient power to maintain altitude, the pilot begins a gentle descent to hold speed. He eases back on the throttle, and diddles with the mixture—the engine is hacking and wheezing, eminently dying, developing power as best as it can—but not enough. The pilot needs to find a place to land, and soon. Leaning out the open window, he scans the ground outside, holding his face to the wind, squinting—straining his eyes to see anything useful. The desert is lighter colored, he can see that—not much vegetation, perhaps only a few trees to worry about, but the pilot watched the terrain carefully while it was yet daylight, and he's studied the map. In a general way, he knows exactly what's down there.…

Rocks. Big, rugged, barren, mountain-sized rocks that would smash his airplane to splinters if he tried to land. He figured 10,000 ft would be sufficient to cross the mountains in the dark, and now the plane was coughing its way through 7500 ft. It might be over soon. There are no lights visible, no roads, no sign of people at all. The pilot can make out the slope of a huge mountain, the details of which can only be imagined. It might be possible to land the struggling plane up the slope, but the plane has no window to the front, is carrying almost 200 gallons of fuel, has no lights, and might not successfully pancake into the mountainside. No telling if the slope there is smooth enough to be useful, anyway. The darkness could easily conceal a menagerie of crevasses, boulders, and petrified logs, any of which could destroy the plane and cause the death of the pilot.

> My mind pictures it—cut by arroyos, spattered with stones, without a level spot where wheels can roll. Of course, I'm lucky to see anything at all—suppose there were no moon, or that the sky were overcast? (Lindbergh 1953, pp. 291–292)

Even worse, perhaps, than dying in the crash, would be surviving for days in the desert afterward. Scorching heat in the day, with freezing temperatures at night, the pilot cringes at the thought of walking out of the wilderness, perhaps injured from the crash, trying somehow to stay alive after being miraculously deposited in darkness, so far from civilization. The ground does not present a favorable situation for landing, so the pilot's attention returns to the airplane.

There is a little time—the engine is trying valiantly to work, slowing the plane's descent. The pilot maneuvers gently to favor a landing on what he hopes is the flattest of the terrain below, and sets to work in the little cockpit. Why would an engine fail? It needs air, fuel, and spark to run. If it were a spark problem, perhaps with one of the magnetos, then switching the mags would make a difference, and there is no change, so it's not spark. Fuel? The pressure is normal, and working the hand pump causes little difference. There is adequate fuel to the engine. Bad fuel? The pilot dismisses that, knowing that the fuel was well strained and the sediment bowl recently drained. Perhaps it is a mixture problem. Too little air would cause the engine to choke on its fuel. Carburetor ice might be blocking the intake, enriching the mixture. Leaning didn't help before, but what's left? The pilot runs the throttle forward to the stop, then back again, hoping to clear the carburetor, and hears a little improvement. Encouraged, he works the throttle and mixture together, carefully, anxiously listening to the sound of the engine as if tuning a musical instrument. It begins to sound better.

Fifteen minutes have passed since the engine's first discordant note. Fifteen minutes of agonizing possibilities, considering a crash in the desert. To the pilot, it seems like hours. The engine runs better at lower altitude—probably because of the warmer air. With the engine running better, the pilot begins a gentle climb back to 10,000 ft and turns toward St. Louis again. There are many more mountains to cross before the plane reaches the gentle flats of the midwest. The cooler air at 10,000 ft will cause the engine to run roughly again, but knowing how to deal with the problem helps. The altitude is necessary to clear the Rocky Mountains. The pilot considers the predicament carefully as he resumes his patient vigil in the cockpit. Carburetor ice. He originally rejected the idea of a carburetor heater on the airplane, in an effort to save a little weight, but he now changes his mind, deciding to have one installed right away, in New York. He'll need it, for Charles Lindbergh intends to tackle the Atlantic ocean in the Spirit of St. Louis (Fig. 2-1).

BACKGROUND INFORMATION

Lindbergh flew through all of one night and part of another on his famous 33 1/2-hour venture from New York to Paris, but that hardly scratched the surface of his night experience. In planning for the solo trip across the Atlantic ocean in a single-engine airplane, he considered the risk no greater than a winter season spent flying the mail. As a St. Louis mail pilot, he demonstrated great skill at flying night VFR, often scud-running above fog and under 300-ft ceilings in near total darkness. The airplanes used to fly the mail were DeHaviland biplanes. Very primitively equipped, by today's standards, the pilots tackled some rather amazing conditions with just their wits and some careful decision making. At the time, pilot licenses had yet to be printed, and charts came from the local drug stores and public information centers. Weather science helped little, was extremely subjective to the particular observer, and occasionally hindered the pilot's task—a national aviation weather service and its standardized reporting system were not yet born. If Lindbergh could see to take off, he'd go. If the weather deteriorated and caught

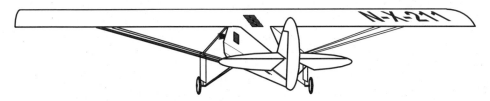

Fig. 2-1. *The Spirit of St. Louis.*

him with no way to an airport, he'd land in a field or a pasture. Moonlight often provided adequate visibility, and if things went really bad, he could always jump—which he did, on at least four occasions after being trapped in the air by fog or otherwise poor visibility. In that event, he would try to run the airplane out of fuel before heaving himself overboard, in an effort to prevent a post-crash fire and thus save the mail. Enlisting the help of some local farmers after one bailout and subsequent crash, he calmly sorted through the wreckage, extracted the bag of mail and sent it on its way to Chicago, safely loaded on a train. Lindbergh often demonstrated his parachute during barnstorming excursions, using a jump over small towns to draw a paying crowd out from their homes to the nearby landing field.

In the famous Spirit of St. Louis, however, a parachute, at almost 20 lb, became dead weight sacrificed for the benefits of longer range; besides, it should be better to ditch the plane in the featureless north Atlantic void than to parachute into the frigid water without a raft. Interestingly, the flight beginning in New York required a plane that was made at the Ryan factory in San Diego, 2500 miles away—over *land*. After a few hours of flight test, Lindbergh made the first leg of the journey nonstop from California to St. Louis—14$\frac{1}{2}$ hours through the dark of night—where a dark desert and a little carburetor ice almost ruined the whole program (Fig. 2-2).

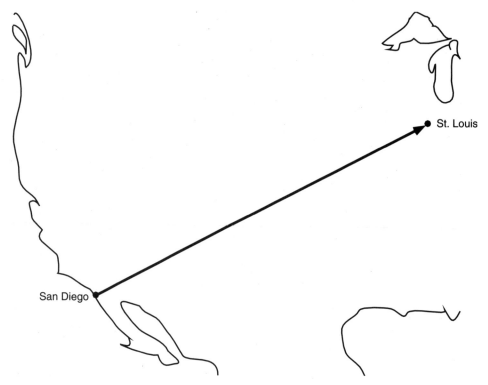

Fig. 2-2. *After St. Louis, he flew nonstop to New York and then…Paris!*

It began with that flight. Then came the journey to Paris. The plane returned triumphant in a ship and proceeded to travel the country for hundreds of hours, visiting almost every major city in the United States, including a lengthy flight nonstop from Washington D.C. to Mexico City. Night and day, engine churning, miles turning, and the airplane flying on. Lindbergh demonstrated, if not yet the practicality, certainly the viability of aviation. He took some calculated risks in the effort. This section will help you weigh some of those risks against practicality, as they relate to nighttime preflight planning.

Engine Failure

To the eye, night equalizes mountains and meadows like ink spilled over the details of a map. If my engine should fail, I'd still have to keep on running, flying,…holding my speed and rhythm to the very instant of the blind, inevitable crash.

Why should I concern myself with engine failure;…I, who am about to take on the entire ocean and the night?…From now on, the explosion in the engine will be inseparable from the beat of my heart. As I trust one, I'll trust the other." (Lindbergh 1953, pp. 290, 292)

Lindbergh considered the Wright Whirlwind to be one of the best available engines for reliability. The factory statistics at the time indicated that operators of that particular engine were experiencing failures averaging only once every 9000 hours—remarkable for the time. So he went charging across the ocean.

Modern engines, particularly turbines, can statistically do much better than Wright Whirlwind, and the facts are that almost any well maintained and correctly operated engine will rarely fail and that when a failure occurs, something *caused* it. The *cause* is the key point of this discussion. What could cause your engine, a complex piece of machinery that is designed to run, to fail? It could be *anything*—and it could happen at *any* time. Perhaps you turned the engine off? Blockage in the fuel? Bad fuel? Lack of air? Fire? No spark? Did some critical part wear thin and jam its inner workings? Did the carburetor swallow a load of ice or the intake ingest a bird? Did the block cool too quickly? Loose bolt? Broken hose? The ever-present cause can be very, *very* small, and you as the pilot will probably not see it coming when it does. So, you have an engine, or perhaps a couple of engines, that probably, almost absolutely, never fail—but they *could*.

"Lucky Lindy" flew the ocean on the hope that his well-maintained and carefully operated engine would run reliably *all the way*. He did so even after experiencing some difficulty with it over Arizona. Carburetor heat should take care of *that* problem if it happens again—but there is no guarantee that the next problem will occur in the carburetor at all. He planned for an engine failure and its subsequent forced landing but did so in an interesting way—to plan too aggressively would render the flight impractical. He had weight limitations—the wings of the plane could carry only so much. Every pound carried in preparation for a disaster equaled fuel that could not be carried to reach the destination.

Chapter Two

Consider it this way: You've got to fly at night and you don't trust the engine. Deep in your heart you sort of know it's going to fail. So you plan the flight to remain over a lighted runway—like Cape Canaveral—all the time, and as high as possible. You insure the plane and your life for enough to make the flight worthwhile. Have a medical team and a crash truck standing by on the ground. Wear a padded, fireproof suit and a helmet. Carry three fire extinguishers. Install "air bags" and extra seat restraints in your plane. Put quick releases and explosive bolts on the doors—better yet, install an ejection seat. Train your ground crew to respond at the slightest sign of imminent failure. Contact mission control. Have helicopters standing by. Foam the runway. Fill out an FAA/NASA reporting form before the flight, so that it only requires a signature. And, on a positive note, arrange for press coverage of your "dramatic" safe landing and serve hors d'oeuvres at the party afterward. You could go overboard. How much preparation is "enough"? Lindbergh considered this carefully.

But suppose my engine fails over the Atlantic, what emergency equipment shall I take with me? Is it wise to carry any equipment for a forced landing on the ocean; or would that simply be a self-deceiving gesture—actually a detriment to safety? Under such conditions, could anything I carry save my life? It's a problem to which I can find no clear-cut answer. Safety at the start of my flight means holding down weight for the take-off. Safety during my flight requires plenty of emergency equipment. Safety at the end of my flight demands an ample reserve of fuel. It's impossible to increase safety at one point without detracting from it at another. I must weigh all these elements in my mind, and attempt to strike some balance. In each instance, I'll try to buoy life with hope, no matter how faint that hope may be. (Lindbergh 1953, pp. 97–98)

In the end, he carried four quarts of water, an Armbrust cup, a hunting knife, four red flares that were sealed in rubber tubes, a match safe with matches, one hack-saw blade, one flashlight, one needle and ball of string, one ball of cord, five cans of Army emergency rations, and an inflatable liferaft with a repair kit. He also dressed in woolen clothing. An engine failure would deposit him in cold water, perhaps thousands of miles away from help. Assuming that he could survive the landing, escape from the sinking airplane, and do so with all of his equipment, he stood a fair chance of floating for a few days. Perhaps a passing ship might see his flare. Hopefully the weather would cooperate—and all this based on the chance that the engine might fail. Modern pilots are no better off. There is *still* the chance that an engine or engines might fail. Are you prepared to survive a forced landing in any of the terrain you might be flying over? We weigh our circumstances in the same balance. To be totally unprepared is perhaps foolish, and being overly prepared could render the flight impractical (Fig. 2-3).

Lindbergh neglected to mention a few vital preparations he made and carried along, which are totally weightless—what of the pilot's skills? If the engine should fail, he knew exactly what a suitable landing site should look like, and he well understood the intricacies of a gliding, dead-stick approach. When was the last time you shot a simulated dead-stick approach to a field? If you fly a multiengine airplane, when did you last prac-

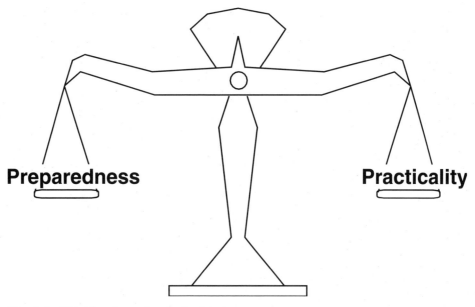

Fig. 2-3. *Weighing the options beforehand...*

tice engine-out procedures? Do you really know how to pick an adequate field; do you have a good idea of the minimum landing distance requirements of your airplane? These skills are vital to your safety in the event of an engine failure, and if you have them, they are no trouble at all to bring on every flight.

Another form of preparation that is weightless in the cockpit is the pilot's choice of route. Wouldn't it be nice if the flight remained over terrain suitable for landing? I have flown single-engine airplanes through mountainous terrain in the dark, carefully following roads, lighted highways and aerial pathways over relatively flat ground. An engine failure under those circumstances would still be a dicey affair, but at least there would be some option in the landing—far better than none at all.

There exists a longstanding debate over whether it's safer to fly a twin-engine airplane at night, as opposed to a single. Fighter pilots take this discussion to extremes, linking it to pride in their particular equipment. Two F-16 pilots happened to be riding in the back of a Boeing 767. Typical for fighter pilots, they looked with disdain at the huge twin-engine jet, flagged down a flight attendant and handed her a note. "Miss, please take this to the Captain." The note said:

Captain—

We'd like you to be aware that you are carrying two of America's finest fighter-pilots aboard your airplane. Yes, we fly the F-16 Fighting Falcon—and, in our opinion, anybody who flies an airplane equipped with more than one engine, is a coward.

The flight attendant soon returned to the young pilots in the cabin, with the captain's response:

Chapter Two

Gentlemen—

Welcome aboard. We're pleased that pilots of your caliber would choose to fly on our airline—and, in our opinion, anybody who flies an airplane equipped with an ejection-seat, is odd.

Fuel

It's 3600 statute miles. The bit of white grocery string under my fingers stretches taut along the coast of North America, bends down over a faded blue ocean, and strikes—about at right angles—the land mass of Europe.... I assumed that the airplane ought to carry fuel for 4000 miles in still air," Hal says. [*Hal designed the Spirit of St. Louis.*] "Maybe that isn't enough. You may want to follow the ship lanes. Suppose you run into a head wind...." (Lindbergh 1953, p. 84)

How much fuel is reasonably enough? A pilot should obviously carry enough to "get there," but what of the reserve? What penalty is owed for carrying too much fuel, and what has darkness got to do with it? In the end, the Spirit of St. Louis landed in Paris with 85 gallons of fuel in its tanks—enough to fly nearly a thousand miles further. Excess fuel lends comfort to pilots fighting headwinds and difficult navigation problems but adds danger to the prospect of an emergency landing. Fuel is rather heavy. Too much of it can hamper flight performance and efficiency, greatly increasing the cost of a given flight, and too little fuel invites an embarrassing engine failure. We're back to the balance again.

Beyond the fuel required to complete a flight to a planned destination, there should be adequate fuel held in reserve for contingencies. This contingency fuel might include enough to circumnavigate hazardous weather, combat headwinds, and perhaps continue to an alternative if landing at the primary airport proves risky. In the night, weather, like the mountains, trees, and rocks below it, in the darkness, and the pilot's options in selecting a suitable landing site become limited to those with artificial lighting. Given the fact that lighted runways are more scarce than not, and that hazardous weather, by virtue of its invisibility, deserves a wider berth at night, the minimum nighttime contingency fuel reserves should exceed those of daylight.

Weather, distance, and terrain are the principal factors affecting the pilot's decisions on fuel. Bad weather may require a more serpentine flight path and greater fuel reserves. Great distances require great amounts of fuel to traverse and expose the pilot to a greater amount of weather hazards. Terrain will dictate the availability of alternative landing sites and the fuel quantities required to reach them. For Lindbergh, a target the size of Europe presented fairly easy navigation, but the sheer size of the distance he covered without the opportunity to land required a large fuel reserve. Thirty-three hours' exposure to an apparently slight 10-mi/hr headwind could stretch the distance the equivalent of another 330 miles. The Spirit of St. Louis cruised at 100 mi/hr, so that 10 mi/hr headwind would lengthen the flight another 3 hours and 20 minutes—so the "headwind" contingency, at normal fuel consumption, could easily create a need for another 50 gallons. For obvious reasons, Lindbergh insisted that he would not depart New York without a tail-wind. For shorter trips, reserves for headwinds may be proportionately smaller.

Weather is the big *unknown* contingency. In daylight, observing isolated storms and circumnavigating them in a relatively straight path to your destination is relatively simple. At night, the easy avoidance of localized bad weather is much more difficult. Compare it to running through a forest: in daylight, an easy task, but a dark night presents an entirely different matter. Occasionally, lightning will indicate the presence of a storm from a great distance; more commonly, however, the airplane is either about to penetrate the system or already inside it when the pilot becomes aware of its presence. The indications may be as simple as the fact that the stars and ground lights appear to "go out" or as alarming as the sudden onset of rain and heavy turbulence. The lack of warning and detail so useful for avoiding weather in daylight forces the pilot to circumnavigate areas of weather at night rather than attempt to pass through them—thus the need for greater contingency reserves for avoiding weather in the dark.

Weather

Lindbergh faced an interesting problem with the weather. The National Weather Service didn't exist for the benefit of pilots. Computers weren't around to help in the briefings. Weather reports carried a high degree of subjectivity, depending upon the opinions and style of the observer. Forecasts depended even more on guesswork and luck than they do today, and there were no reporting stations along Lindbergh's intended route of flight, which crossed the sea well north of the shipping lanes. He flew into the unknown with an airplane minimally equipped for instrument flight, unable to combat icing, unable to detect weather invisible to the eye. The plane carried no lights, no radio, no LORAN, no GPS, no stormscope, no radar, no boots, and no forward windscreen. A few times, its brave pilot realized he'd flown into a rain shower by nothing more than the pin-pricks of rain drops striking his hand, extended out the window. Lindbergh departed on the hope that the best weather scientist in New York would guess correctly—that the weather over the North Atlantic was clearing. Halfway across the ocean, that weather forecast would already be 24 hours old.

There were advantages in the New York/Paris flight. There are few mountains in the open sea. Lindbergh could be fairly confident that any altitude above sea level would suffice to clear obstacles. A ship, perhaps, might obstruct the flight path, but he flew far from the shipping lanes. Birds? Maybe, but they presented no greater hazard than on his overland flights. Other airplanes? No! That's one of the benefits of being the first. So he could fly under weather, within a few feet of the waves if necessary, and around it for thousands of miles in any direction—as long as fuel would allow. The wind left tell-tale signs on the water, making strength and direction estimates possible, and navigation consisted primarily of flying consistent headings on the compass.

He did encounter weather for much of the flight, including ice, rain, fog, and thunderstorms. He flew on instruments for extended periods and for a long night above the clouds. Skimming the ocean waves to avoid weather became unthinkable in the darkness below an overcast. It was hard to write in the log—he was tired, and holding a flashlight, flying the plane, and writing at the same time felt like an impossible task. When you put it all together, that flight across the sea against unknown weather required remarkable courage.

Modern weather reports have the benefit of a standardized reporting system. Weather observers are certified and approximate an accepted standard in their reports. Computers collect and disseminate the data, drawing on huge files of past weather history, allowing meteorologists to forecast a little into the future. How trusty are the forecasts? They are still a best *guess.* Weather is consistent only in its ability to change—so change has to be assumed. Weather *reports,* by virtue of the fact that they indicate *present conditions,* can be accepted as truthful; however, a weather report, as soon as it is created, becomes a picture of conditions as they *were* (even if only a minute ago)—and some change must be assumed. Thus, the most recent weather reports carry the most weight in evaluating current conditions, with older reports fading more and more into the realm of history, useful primarily for establishing a trend.

For a pilot launching into invisible weather conditions—flying at night—weather reports provide a reliable indication of the weather as it most recently *was.* Pilots may hope that whatever changes have occurred in the weather since the reports have been made are inconsequential, but prudence dictates that they prepare for something worse. If the reports call for clear skies, the pilot expects that or possibly some clouds; scattered clouds might foreshadow an overcast; IFR conditions might imply storms or more serious hazards—then again, the weather could improve. The scary point of the matter is this: Whatever changes have occurred in the weather at night are unknown—until a new report is issued. This is different from daylight conditions where a pilot makes constant visual observations of the weather during the entire flight. This is not to say that the night pilot is entirely blind. Nighttime weather is often detectable in ambient city lights or moonlight. Sometimes the strobes or landing lights will indicate the presence of a cloud. Turbulence does not manifest itself visually anyway, and lightning is far more visible in darkness than it is in daylight. Nevertheless, without the benefit of visual conditions, there is far less weather information available to a pilot flying at night—and a report alone is nothing to base your life on.

With this in mind, most careful VFR night pilots prepare for unexpected IFR conditions and raise the minimum night VFR weather requirements far above their daytime standards. If the pilot plans to tackle night IFR, then the inclusion of certain weather detection equipment is desirable, for example, Stormscope, Strikefinder, or radar, useful in detecting the unreported or unforecast storm or weather hazard.

Light

I've known only a few people who pay close attention to the phases of the moon. Most had a scientific interest, and the rest were simply weird. Night pilots need to watch the moon carefully—it is the second best source of light available. All that worry about forced landings in the darkness could be abated by bright moonlight—you might even get to see and avoid a lot of the weather en route.

As big an advantage that moonlight provides, it's absence could really leave you up in the dark. Often, moonlight only lasts through part of the night, or rises only after a long period of darkness after sunset. The light of a partial moon is much less than when it is full, and when the moon is close to the horizon, its light is much fainter than when it's high in the night sky. If the time of a night flight is somewhat flexible, the pilot might greatly multiply emergency options and increase safety by taking off when the light of the moon is brightest—howling at the moon is optional.

City lights provide some interesting options at night—one of which is the pilot's choice of emergency landing fields. Some areas which are exceedingly poor runway choices in daylight become a better, and often only, option at night. Parking lots, for example. Although many are large enough to serve as a runway (like shopping malls) by day, they are strewn with cars, people, and noisy activity and are very unlikely to be an appropriate destination for a quiet airplane that is committed to land (Fig. 2-4*a*). At night, these places change. Most of the cars and people go home. The lights come on—okay, I know the light posts and wires present a difficult problem, but at least the pilot can see them—making a large, lighted, relatively obstacle free area in which to crash-land an airplane. (See Fig. 2-4*b*.)

Conversely, city lights can work against the moonlight. Terrain such as open fields and unlit runways that are adequately visible in moonlight alone become black holes if they are surrounded by city lights. That's because your eye adjusts for the brightness of the city lights and turns down the dimmer reflections cast by the moon. Perfectly useful landing fields in the midst of cities become almost impossible at night—even in bright moonlight. Get far enough away from town, however, and the moonlight reigns supreme,

Fig. 2-4a. *Parking lot by day.*

Fig. 2-4b. *Parking lot by night. It won't be easy, but it's better than the alternative.*

making an open-field landing possible again. The pilot can also hope for fewer wires, poles, and houses as the plane glides away from the city—never mind possibilities of the occasional horse or herd of cattle—night emergency landings will never be easy.

Terrain

With the emergency landing in mind, the terrain over which the pilot plans to fly takes on a new twist. A friend of mine flew his Cessna 150 from Salt Lake City, Utah, to Phoenix, Arizona, one evening and arrived tired, well after dark. Trevor was impressed by the lights of the big city, and wondered what it would look like in the daytime. There appeared to be large areas of water down there among the lights. Big dark shapes that he assumed were lakes. (He didn't think that maybe Phoenix was in a desert.) He flew all around the city, fairly low, looking happily at the dazzling scenery below, passed close to one of the "lakes," and landed at the airport. When he awoke the next morning, he startled to the fact that on landing, he'd passed within a few feet of a small, but rugged, mountain. Other mountains just like it were scattered about the area like—*just* like—the small "lakes" he'd seen the night before.

It might be wise to examine the topography of the area in which you plan to fly. Look closely for terrain, obstructions, and unlit obstacles like antennae, buildings, and smokestacks and pay very careful attention to the area around the airport at which you intend to land. You never know when you might be surprised by a "lake." (See Fig. 2-5.)

Navigation

Far and away, the most accurate and practical navigation method is pilotage. When you carefully guide food from your plate, onto your fork, and into your mouth, you employ a form of pilotage. So accurate is this technique of visual guidance that you often don't need the napkin. Driving a car is accomplished entirely by pilotage, which means that you see the road and manipulate the controls to follow it. You can parallel park in heavy traffic this way. In the cockpit, pilotage will be used for at least part of every flight, not limited to taxi, takeoff, and landing, where the pilot visually guides the airplane with reference to its position over the ground. This is accurate, precise, and completely irreplaceable—which is why there are no blind pilots.

Blind flying, however, is possible by limiting the pilot's attention to the gyroscopic, magnetic, and electronic instruments on the panel. It has certain inaccuracies; heavy, congested traffic, for example, occurs when the individual airplanes pass laterally *within miles* of each other, and vertically *within 500 ft*. I'd love to see congested interstate highways with spacing like that. The large spacing requirements are necessary because of the inadequacies of instrument flying, as opposed to visual separation—where the pilots can see the other planes outside, like the air races at Reno. Even an instrument approach, using an instrument landing system (ILS), the most capable of all, can be hand flown, at it's best, no closer than *200 ft* from the ground, unless the approach environment is acquired visually. The key word is, of course, *visually.*

Fig. 2-5. *Looking south along the Wasatch front, from 19,000 ft over Brigham City. The dark void on the left is mountains, and the dark area on the right is part of the Great Salt Lake.*

Chapter Two

Electronic navigation offers the pilot a direction to fly, following instruments, or a backup position indication when flying visually. The most recent developments in this field are satellite-based global positioning systems (GPS), and still much less capable than eyeballs alone in close maneuvering—especially when the satellite signals are downgraded for civilian use. High-frequency radio navigation and all the other aids accomplish the same purpose with even less accuracy. They are the best available and still not as capable as your own eyeballs.

Dead reakoning, though the simplest of navigation principles, has the largest margin of error and yet is easily corrected with a map and a good look at the ground.

Of all forms of navigation, only pilotage will allow you to fly down a mountain canyon, close above the ocean waves, or land on a runway. As you lay out the course you intend to fly in the dark, you must consider the inaccuracies of the navigation method you intend to use and prepare for those margins of error to increase. If you plot a visual course through a mountain pass, for example, and encounter unexpected weather in the middle of it, suddenly your margin for error is unexpectedly tight. It would be frighteningly difficult to avoid the rocks while flying blind and using dead reakoning alone. If you expect weather, the best choice is also to expect some instrument conditions. Planning for instrument conditions means selecting a VFR route and altitude compatible with IFR procedures, i.e., sufficiently high to clear all obstructions in the path with a wide margin and within useful range of ground-based navigational stations. If the weather deteriorates at that point, you would be in a far better position to transition to an IFR flight plan and continue the flight.

Rest

A soldier fresh from interrogator school made this comment: "Give me any prisoner for a week and I'll get anything out of him, *anything*." I didn't want to think about what the sadistic pervert would use to do it, but as he continued, my opinion changed. "I'd just keep him awake." That was it. If you're denied the opportunity to sleep for much longer than a few days, you break completely; becoming incoherent, hallucinating, sleeping with your eyes open, losing control of your bodily functions, and getting physically sick. You will plead for sleep, and if you happen to be flying an airplane, you may become completely incapacitated and crash.

For Lindbergh, unexpected good weather surprised him with a sudden opportunity for departure. In the excitement that followed, he received less than two hours of sleep the night before. By the time he landed in Paris, he'd been awake for more than 54 hours. His desire for sleep *during* the flight was intense, particularly during the night hours, and presented his single greatest challenge—that of simply staying awake. He had visions, heard voices, and often caught himself dozing off as the airplane spilled off its heading. He spent many hours in a kind of stupor. He slapped his face, hard, and felt nothing. His arms and feet became heavy and sluggish. He didn't want to fly anymore, he didn't want to do anything but *sleep*. He employed remarkable willpower in the face of this terrifying prospect: "You fall asleep, you die."

You intend to fly at night, so you risk the same hazards. Especially if you've been awake all day before. You can expect your skills to be dulled, your mind to be slower, and your judgment to be adversely affected. So get adequate rest before you go, and avoid it all.

Equipment

What to take with you on a night flight is an open-ended question and subject to debate. You're looking at the same balance that Charles Lindbergh did in his flight across the ocean. Here are a few recommendations:

An airplane, obviously, but make sure it's in good working order. If you're unsure about the engine or if the plane's recently out of overhaul or heavy maintenance, find out if the engine is okay in daylight. Test flying at night seems to be inviting trouble. Here, I suppose, is a good opportunity to describe the minimum airplane that is truly up to the task, *really,* of flying many hours at night. The night pilot needs to maintain a few options in the face of an emergency and a single-engine does not provide many when the engine quits. Start with more than one engine. From there, consider the terrain. Is it higher than your light twin's single-engine ceiling? If so, get a more powerful airplane, preferably one that can maintain the IFR minimum en route altitude on just one engine, at any point in the flight. If you're flying in the dark for very long, you will encounter unexpected weather. It would be nice to see it coming, so radar and passive weather detection equipment (Stormscope, Strikefinder, etc.) are a good place to start. Finally, since you may encounter unexpected weather and that weather might contain some ice, it'd be good to have anti-ice or deice capability. That's just the airplane. If the equipment is available, I'd hope that the pilot knows how to make good use of it.

The foregoing is a wish list. In spite of all that, and all the money it takes to make it go, single-engine (or even less!) night flight is still possible to accomplish with an acceptable degree of safety. The single (or glider) in most cases has a strong advantage in a forced landing because of a generally lighter wing loading and better controllability, and many night accidents are simply controlled flight into unseen terrain, where multiple engines are of little help. I might suggest that the VFR, single-engine pilot pays closer attention to the area below the plane, enabling a little better situational awareness when a problem arises (or descends—as the case may be).

Bring a flashlight. It's mighty hard to read the topography of an aeronautical chart while you're trying to hold it up to the city lights outside, and if the cockpit lights go out, you don't want to be caught wishing you had taken up smoking. What kind of flashlight to bring is up to you—I know one very dedicated pilot who always brings three: one on his keychain, one in the bag, and one on his head (he fancied the spelunkers' creed of three sources of light). A bright flashlight could destroy your night vision, so one with a dimming adjustment is handy. Red light does not seem to affect your ability to see in the dark, so a red filter is also useful. Bulky flashlights often have expensive batteries and are difficult to hold in your teeth, should you ever need a free hand. There are literally thousands of options. The airline I work for requires each member of the flight crew to carry a flashlight that is powered by two D-cell batteries, according to FAA rules. Find the one that is right for you, and don't leave for the airport without it. So carry at least one flashlight (Fig. 2-6).

Fig. 2-6. *Selection of the proper flashlight is a personal decision.*

A chemical source of light can come in handy if your flashlight doesn't work. This is often a self-contained gizmo that consists of two reactive chemicals, one isolated in a glass vial, inside a flexible plastic tube. You bend the tube until the glass vial inside breaks, allowing the two chemicals to mix, shake the whole thing, and it glows like a hyperactive firefly for up to 8 hours. Often sold under the name Cyalume light-stick, it is inexpensively available at your local hardware store. Since these have an almost unlimited shelf life, they are easy to keep in the bottom of your flight bag, and they're cheaper than an extra set of batteries.

The FAA says you have to have at least position lights on your airplane. If they are inoperable, you can't fly. Depending upon the regulations you operate under, landing lights might also be required. During the preflight inspection, you'd better energize the electrical system and walk around to inspect the lights—otherwise, you could be looking at a violation.

Make sure the instrument panel lights work on the airplane. Believe it or not, these are not required by the government, but flying by the light of a flashlight alone can become quite tiresome, and it wears out your batteries.

Bring an operable aviation band radio (Fig. 2-7). Most unattended airports use pilot-activated lighting systems. Key the microphone a few times and a wonderland of runway and taxi lights appear. Without a radio, you land in darkness. If you don't trust the radio

Fig. 2-7. *You'll need an operable radio of some kind to turn on the runway lights.*

in your plane, carry a spare one or enough fuel to continue flying to an airport that is continuously occupied.

A means of basic IFR flight could save your life in the dark. This does not necessarily mean that your panel should resemble that of the Space Shuttle, but at least carry a couple of gyros in good working order. Lindbergh had the equivalent of a turn and bank indicator and a compass, and if his turn gyro ever failed, he would have been soggy toast. Most general aviation airplanes have attitude indicators *and* a turn gyro, one electrically powered and the others by the vacuum system. This is adequate for basic IFR. You need some direct or indirect attitude information and an idea of your heading. VOR, GPS, LORAN, ADF, DME, RNAV, FMS, INS, IRS, and all the other electronic goodies on your panel are useful for navigation, but basic IFR simply requires the ability to keep the airplane upright—an impossible task with radio navigation instruments alone.

The FAA recommends the use of an oxygen system for night flight, suggesting the pilot employ it at altitudes above 5000 ft. That's below pattern altitude at most airports in Utah, and, aside from the fighter pilots, it would be hard to find a pilot who makes regular use of oxygen around here. (Maybe if we could wear a fighter-pilot helmet, too?) This depends entirely upon your personal needs. Oxygen does help you see better in the dark, and we'll talk about it later.

Bring a little survival gear and include a little first aid equipment. Since the odds of emerging from a forced landing at night and unhurt are poor, you'll need to help yourself until help arrives. Should you go down, the search will wait for daylight to start looking for you. It might be painful. First aid should include a little something for bumps, scrapes, broken bones, bleeding, burns, infection, and shock. Nights are usually cooler than daytime, so bring appropriate clothing and perhaps include something to start a fire—the magnifying glass included in lightweight survival kits will not work in the dark. Consider the types of terrain you're flying over; what would be the nature of the emergency landing in that terrain? If you fly over forests, landing in water might be the best choice, so bring a raft or life jacket. If you fly over the desert, bring some water. The options here are endless. Use your head, weigh the balance, and make the best of it.

There have been a few popular aviation writers who wouldn't dare fly single-engine at night or over a low overcast without a parachute. This is a hairy option, but it worked well for Mr. Lindbergh, and it may be an excellent choice if your plane has quick release doors and only a few professional occupants—namely you, alone. If there is company in the cabin, and you really want an airborne option if the engine quits, you should consider the multiengine hardware. Parachutes just don't sit well with a crowd. I can just see it: You're flying along in your late model Piper Malibu, the engine throws a rod and quits, undeniably and catastrophically. The white-faced, terrified business executives in back stare unblinkingly at you, the pilot, while you calmly (?) announce that when the cabin door seal finally deflates, you will try to open the door and "please put on your 'chutes and follow me outside." Hopefully you're still high enough for the 'chutes to be effective, and then who knows what you'd land on. Maybe your passengers would be thrilled to jump out, but I wouldn't count on it—you'd have better luck getting Congress to lower taxes.

I often wonder if night vision goggles would be helpful. Having never used them, I posed the question to a couple of soldiers and helicopter pilots. They weren't sure, but it seems like a better option than the parachutes.

TECHNIQUE

So there you sit, contemplating a night cross-country flight. Ignoring your reasons for attempting it, we'll look at a few methods you might use to plan the flight that would be different from one in daylight. Assume that your plane is adequate for the job; equipped for light instrument flight and in good working order. The flight will begin and end in darkness and you plan to go VFR, hoping to enjoy some of the delightful scenery created by the lights below. Begin by laying out a good VFR chart.

The Chart

The chart is your only real picture of the terrain. If you plan to fly relatively low, that is, less than 10,000 ft, use the standard sectional VFR chart to plot the course. World aeronautical charts (WAC) are nice for faster airplanes because you do not have to change from one chart to the next as often, but they ignore much useful detail. If your trip is lengthy, assemble the various charts together, tape the edges, and fold the composite assembled chart around your courseline. You should have one long folded strip of chart, about a foot wide, with your courseline running down the middle. Fold it again, like an accordion, and stick it under a clipboard or throw a big rubber-band around it. This goes with you in the cockpit and you'll be referring to it in real time. (See Fig. 2-8.)

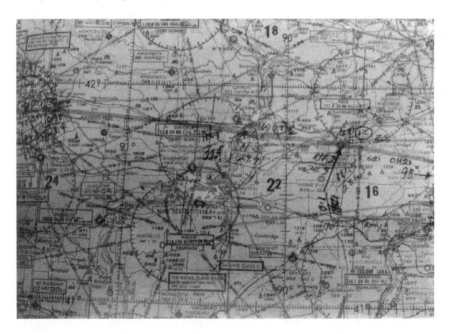

Fig. 2-8. *A chart is your best navigational resource.*

Your course line should be marked brightly in a color you can see (green is invisible in red light). Use a highlighter pen of some type so that you may yet read the data underneath. Do away with the flight planning log and write directly on the chart. These things are superseded every 6 months anyway, so it's unlikely that the chart will become messy with writing—you'll just get a new one. Writing directly on the chart helps you focus on the information there, instead of referring to yet another source or loose paper in the cockpit. Mark your checkpoints and calculated times and fuel burns in appropriate places along the course line, and write the actual data next to them as you fly. In this manner, your attention is focused on the relationship between the ground outside and the chart inside, making your pilotage and navigation that much easier.

If you suspect IFR conditions en route, it's a good idea to have the appropriate IFR charts already selected and handy. Night instrument conditions often happen unexpectedly, and on those occasions you'll already be in the clouds before you realize that VFR is no longer possible. Under those circumstances, you'd like to be ready for the transition to IFR procedures, not hunting at length around your flight bag for an en route chart while the airplane wanders away from you.

Airports and Facility Information

Figure 2-9 shows what information about airport facilities you get from the sectional chart. It's quite a bit, really, but you need to know a little more.

You need lights at both airports you intend to use. The charts indicate the presence of a lighting system, but you'd better check with flight service to see if any parts of them are broken. You probably need fuel, too. What time does the fixed base operation (FBO) close up and go home? Do they have a call-out fee? Perhaps you could avoid some expense by fueling your airplane in the daytime. This information is often carried and updated in the Airport/Facility Directory (AFD) or the more popular (and much more expensive) books such as *Flight Guide* and *AOPA Airports USA*. The AFD's purpose is to

> contain data on airports, seaplane bases, heliports, NAVAIDS, communications data, weather data sources, airspace, special notices, and operational procedures....Including data that cannot be readily depicted in graphic form; e.g., airport hours of operation, types of fuel available, runway widths, lighting codes, etc. The Airport/Facility Directory also provides a means for pilots to update visual charts between edition dates; i.e., the [AFD] is published every 56 days while the Sectionals and Terminal Area Charts are generally revised each six months." (NOAA catalog, 1996–1998)

See? You're *supposed* to write on the charts, and you should include all information pertinent to lighting and runway numbers that would be necessary to you in the cockpit. The phone numbers, FBO names, and all the rest might be left in the book for referencing after you land.

You're trying to avoid circling the destination airport's beacon, keying the microphone on every frequency you can think of, unable to turn on some runway lights, and, if you do turn them on, guessing what the runway number designation is. It's best to have

Fig. 2-9. *Night flight requires additional information that is not contained on the charts.*

that information down before you get there. You won't be able to look at a painted runway number in the dark.

For a free copy of the current National Oceanic and Atmospheric Administration (NOAA) aeronautical chart catalog, write or call:
NOAA/National Ocean Service

Chart Sales Office
6501 Lafayette Avenue
Riverdale, Maryland 20737-1199
1-800-638-8972

Route

Plan your route carefully. Look at the terrain over which you'll be flying. Are you willing to land there? There is always the possibility of making an emergency landing anywhere along the route. You're flying at night, so the options need to be pretty obvious. You might avoid broad areas of trees, mountainous terrain, and water, choosing instead to follow roadways and populated areas. There are more lights along the roads anyway, and the night scenery will be better. If the moon is bright, you might consider traversing open areas away from roads, hoping that moonlight might be suitable for a landing there. If you're flying a twin, you may safely plot a straighter course, except keep in mind the limits of your airplane and plan to avoid any terrain that rises higher than you can fly with one engine caged.

You'll need to pay attention to the weather as you plan the route. Reported weather along your flight path might be cause to plan the course around it. If you are IFR-capable, and willing to do it, plan the flight to follow close enough to the airway system to make the transition from VFR to IFR relatively simple.

Weather

Get detailed weather information from your computer or flight service. Try to picture location as carefully as you can and mark it on the chart, along with everything else. You might jot the basics of the destination weather down on the chart as well—there might not be someone awake to give an advisory when you arrive. If an in-flight weather advisory or an in-flight weather advisory of serious nature (AIRMETS or SIGMETS) exists for your route of flight, you could box those areas in, on the chart, as they are described in the verbal reports, and have a spooky picture to look at. This would give you a nice idea of where to expect things like mountain obscuration, thunderstorms, turbulence, IFR conditions and icing, just to name a few. If it looks bad, you might reconsider your planned route and fly around the bad weather.

The weather will be the primary reason that you'll divert to an alternative airport. If the weather looks marginal at your destination, now is a good time to consider the options. Try to pick an alternative (or two) that lies outside the forecast area of bad weather.

Flight Following

Flight following is a service provided to VFR pilots only on a workload-permitting basis. If the controller is too busy to assign a transponder code and track your flight, then you may do it yourself with regular position reports to flight service. Among the many advantages of flight following is that your position is known to the controller—you have been radar-identified. Should you encounter unexpected weather and need to transition quickly to IFR, the controller will be able to help you smooth the transition if he or she already knows where you are.

Flight plans are like telling your mom where you're headed. She knows what time to expect you and whether to worry if you are late. The flight service station (FSS) is like your mom. So file a flight plan for all your night cross-country flights. If bad weather is expected and you plan to use the IFR system for part of the trip, figure approximately where the transition will occur and file a composite flight plan. If the weather is better than you expected, simply continue VFR and don't activate the IFR segment.

Performance and Fuel

Now that you have compiled all the pertinent information for the trip, it's time to calculate the required fuel. As was discussed earlier, plan to carry enough fuel to get there, continue to an alternative, and then a little reserve. It's common sense. Mark your expected position and fuel consumption at various points along the chart and log the actual figures alongside for comparison as you fly. If the planned numbers come up substantially optimistic, you might rethink your options en route, perhaps considering a stop for fuel.

In another book, *Piloting for Maximum Performance* (McGraw-Hill, 1996), I suggested that you keep a running log of all your cross-country flights in a binder so that you might get a trend analysis going. This is the best way to predict the performance of your airplane and the recommendation applies here, especially if you've been on this same trip before. This trip log is something that should take none of your time during the flight—you should be using the chart—but you should take a few minutes to record the data afterward.

Physical Preparation

Fatigue and vision are the single biggest physiological factors of flying in the dark. You can do much to alleviate the fatigue by getting adequate rest beforehand. You may still be sleepy, however, simply because it is night, and people generally sleep at this time. Some things that may help you stay alert in the cockpit are, among others, active pilotage and looking for emergency landing areas, chewing noisy food, exercising in your seat, and sucking on ice-cubes. In planning the flight, you've already made the preparations for the pilotage part; obviously, chewing noisy food and ice cubes will require that you bring some food along. I like to eat carrots, corn-nuts, chips, or whatever makes a strong crunch when chewed. It works wonders for staying awake. The ice cubes are nice to suck on and if you fall asleep with one in your mouth, you'll choke (wake up) or dribble it out on yourself—it's cold—and you'll wake up. Whatever you choose, light noisy snacks are indeed helpful in staying alert. It works in your car, as well.

There are several isometric exercises that can be accomplished while sitting at the aircraft controls, and we'll discuss these in a later chapter as they require little advance preparation.

SKILLS TO PRACTICE

One of the best life-saving skills to possess in flying, whether day or night, is a sense of distrust. A suspicious pilot finds it difficult to become complacent. I'm not advocating paranoia here, just that a pilot's mind should be continuously thinking of possibilities, weighing options. On a cross-country flight, for example, a careful pilot often studies the terrain below, looking for suitable landing sites to use if the engine should fail. The same pilot listens carefully to engine noise and considers every small change in the instrument indications, looking for *anything* that could become a problem. This pilot is always trying to see it coming, whatever it is, and be a little prepared before the emergency happens (Fig. 2-10). A pilot who does not possess this attitude finds it easy to fall asleep in the cockpit.

With this in mind, there are a few skills that could be very useful, if you work to make them habitual. The first is to study the ground. This is fun. One of the reasons you began flying in the first place was for the incredible view, wasn't it? So look at it. Look for the little details which give away the direction the wind blows, subtle slopes in fields and wires across roads. Find areas that might be easy on your wheels and long enough to be useful. Consider some of these same places in the dark. If you have to dodge a couple of trees to use the field, you'll more than likely hit them in the night. Are there animals

Fig. 2-10. *Considering the possibilities...*

grazing down there? *You think they'll run out of the way?* That's unlikely even in the day-time. Almost every cow that I've buzzed jumps and runs some time *after* the fact, deer seem to bolt *into* the flightpath of the airplane, and, in the dark, the cattle will probably be sleeping. You need really good fields for a night landing. They need to be much larger than the minimum daylight field and much easier to approach. Look hard enough and you'll find a few that will do nicely in a pinch. The key here is to be constantly looking—that's skill number one.

The second skill to practice will be referred to frequently throughout this book; that is, brush up on your basic instrument flying. If you aren't instrument-rated, get an in-structor and go practice—at least to the point that you may confidently maneuver the air-plane by reference to instruments without losing control of it. Instrument flying procedures themselves are a topic well covered in many excellent books on the subject and will not be treated with too much detail here. Avoiding all the procedural complexi-ties of the IFR system, simple attitude instrument flying skills with the ability to go from visual conditions to instruments and back rather quickly should be sufficient for most VFR night challenges—assuming the weather doesn't go really bad. The practice is gen-erally interesting, and the skills you will obtain are invaluable in facing the challenge of night flight. By the way, that's all Lindbergh had—approaches, ATC, Jeppeson/NOS charts, and the airway system as it is did not yet exist.

Finally, prepare for an engine failure. Hire a good instructor and go out there, pull the throttle back to simulate the problem, and attempt to do something about it. For mul-tiengine pilots, this means maintaining control, reconfiguring the airplane, adjusting trims—and possibly making an emergency landing in the terrain below—that second engine does not always guarantee a runway. You'll need to make a decision as to where land or to modify your course, as your flight probably will not continue as planned. Single-engine pilots get to make an immediate approach to one of the landing areas they had already picked out (you were looking, weren't you?). You should be able to glide accurately into a position from which the field can be safely used, having the airplane configured for an emergency landing. These skills should be practiced at night as well as daylight and considered at any point during the flight as well. Be careful. You can guarantee that whenever the engine failure occurs, it will be at a most inconvenient time. When you can confidently deal with engine failures, and you are well practiced in the flight skills and decision making involved, you will be that much better prepared when it happens *for real.*

OK, you've planned your night flight carefully and, on paper at least, feel that you can do it with an acceptable risk; but what about the legalities?

FURTHER READING

Lindbergh, Charles, *The Spirit of St. Louis,* Charles Scribner's Sons, New York, 1953.

3
Committing aviation in the dark

It looked like a UFO, bringing crop circles and laser-beams from some distant star. The two farm boys stared across the cornfield, barefoot and open-mouthed, from the shelter of an oak tree. Bright, blinding light flashed low over the corn, as mist trailed out from under the craft. It flew the length of the field, close enough to smell it, then darted high at the end, turning, with bright lights shining sideways for a moment, then off. It flew back across the field, going the other way. The boys weren't impressed.

"I think Harold flies lower in the daytime," said the older boy.

"Yeah, I can't hear his wheels hitting the stalks, but it sure looks cool—like a UFO or somthin'."

"Harold ain't no UFO. Leastwise, he's got the same plane as always—just lights it up at night—and UFOs don't sound like dusting planes with propellers, and all."

"Since when do you know all about UFOs? Maybe Harold really is an *alien*—comes out at night like some kind of mosquiter bug, an' sucks blood out of our cows." The younger boy spoke like a kid around a campfire, eyes wide in the dark.

"Shut up. Maybe *you're* the alien....Say, how do you suppose Harold misses them wires in the dark?"

Chapter Three

The fascinated boys watched as the plane swathed back and forth across the cornfield, darting over and under the invisible wires, and dodging fences. Coming at them, it blazed light, making the cornfield brighten like a stadium at night. Although the noise it made shook the air around them, the boys couldn't see much of the plane, only a bright blur of the propeller, a bit of the engine and cowling, and dimmer reflections from the wings—it resembled a huge noisy moth, swooping and darting, pushing bright lights ahead, like something from another world.

"Do you suppose that's legal?"

BACKGROUND INFORMATION

What's legal in the dark? Almost anything. You can fly formation, aerobatics, go crop dusting and tow banners and gliders at night, with virtually identical restrictions as in the daytime. Crop dusters commonly stretch the flight hours late into the night, during the prime work season. Many claim that steel wires are reflective in their floodlights and easier to see at night, winds are generally calmer, and the air is often smooth.

Thankfully, the United States has perhaps the most lenient night regulations, and surprisingly, most of the rules currently in the books pertaining to night flight actually make sense. The following is by no means a complete listing of the Federal Aviation Regulations (FARs)—just a look at some FAA rules that have direct bearing on the night pilot. We'll start with some basic definitions, and to avoid the sleeping sickness that occurs when reading government regulations, I'll throw in a comment or two for your sanity. FAR part 1 gives definitions and abbreviations.

FAR 1.1 Definitions
Night means the time between the end of evening civil twilight and the beginning of morning civil twilight, as published in the American Air Almanac, converted to local time.

The official government stance on the question "What time is it anyway?" is answered basically like this—if you step outside, can't see the sun, and it looks pretty dark, it's officially nighttime. You should bring a flashlight and at least put on your shoes.

Controlled airspace means an airspace of defined dimensions within which air traffic control service is provided to IFR flights and to VFR flights in accordance with the airspace classification. NOTE—Controlled airspace is a generic term that covers Class A, Class B, Class C, Class D, and Class E airspace.

"Air traffic control service" means that a controller assists the pilots in maintaining aircraft separation. "Controlled airspace," in itself, does not imply the existence or participation of a controller—for pilots, it indicates areas of expected high traffic where weather minimums are relatively stringent, as compared to "uncontrolled" airspace. In theory, higher weather minimums should facilitate the "see and avoid" concept and prevent midair collisions.

You could draw a comparison to public versus private roadways. On public roads, a speed limit exists which traffic is expected to obey, regardless of whether the drivers are on a patrol officer's radar. No such limit exists on private roads, as fewer cars traverse them. Nevertheless, rules exist in common for both—it is illegal to drive drunk on both public and private roads, for example. Controlled and uncontrolled airspace share the same relationship. Controlled airspace can be expected to carry higher traffic but not necessarily radar control. Uncontrolled airspace has more lenient weather minimums (the chance of a midair is less) but still shares many common operating rules. Boiled down, the differences between the two are primarily in their respective weather minimums.

> *Flight visibility* means the average forward horizontal distance, from the cockpit of an aircraft in flight, at which prominent unlighted objects may be seen and identified by day and prominent lighted objects may be seen and identified by night.

If the time is "night" and the airspace "controlled," the government may require more flight visibility.

> *IFR conditions* means weather conditions below the minimum for flight under visual flight rules.

Nighttime, in many areas, would qualify as IFR under this definition. Interestingly, the VFR weather minimums in uncontrolled airspace would qualify as definite IFR in most controlled areas.

> *Instrument* means a device using an internal mechanism to show visually or aurally the attitude, altitude, or operation of an aircraft or aircraft part. It includes electronic devices for automatically controlling an aircraft in flight.

Attitude instruments should be a part of night flight operations.

> *Nonprecision approach procedure* means a standard instrument approach procedure in which no electronic glide slope is provided.

This would also include the typical night landing.

> *Operational control,* with respect to a flight means the exercise of authority over initiating, conducting or terminating a flight.

That's you, the pilot; unless you work for an airline, in which case you get to share the responsibility with your dispatcher and your boss. Once airborne, though, it's *your* butt on the line.

> *Pilotage* means navigation by visual reference to landmarks.

This may prove difficult at night.

> *Pilot in command* means the pilot responsible for the operations and safety of an aircraft during flight time.

That's you, again—this helps the government know who to blame.

> *Precision approach procedure* means a standard instrument approach procedure in which an electronic glide slope is provided, such as ILS and PAR.

This information can be very helpful on a night landing. In part 121, Operations, it is illegal to make an approach below an operable ILS or PAR glide slope—if one happens to be available to your runway. Visual approach guidance, such as VASI or PAPI lights, are also regulatory by nature, i.e., it is illegal to approach the runway below their indicated glideslope at night.

FAR 1.2 Abbreviations and Symbols

VFR means visual flight rules.

Again, "night" is not necessarily VFR.

REQUIREMENTS FOR PILOT CERTIFICATION—NIGHT OPERATIONS

FAR part 61 lists requirements for night operations experience.

FAR 61.57 (b) Night Takeoff and Landing Experience

(1) Except as provided in paragraph (e) airline operations, of this section, no person may act as pilot in command of an aircraft carrying passengers during the period beginning 1 hour after sunset and ending 1 hour before sunrise, unless within the preceding 90 days that person has made at least three takeoffs and three landings to a full stop during the period beginning 1 hour after sunset and ending 1 hour before sunrise, and—
   (i) That person acted as sole manipulator of the flight controls; and
   (ii) The required takeoffs and landings were performed in an aircraft of the same category, class, and type (if a type rating is required).

This rule was recently changed to stipulate that the pilot be sole manipulator of the controls. That is not a requirement while the pilot is in training (thank goodness), as you'll see later on.

Airline transport certificates carry IFR privileges. All scheduled airline flying (except for scenic flights) is conducted according to IFR procedures. Airlines have their own recency of experience, check rides, and certification requirements.

FAR 61.87 Solo Requirements for Student Pilots

(m) Limitations on student pilots operating an aircraft in solo flight at night. A student pilot may not operate an aircraft in solo flight at night unless that student pilot has received:
   (1) Flight training at night on night flying procedures that includes takeoffs, approaches, landings, and go-arounds at night at the airport where the solo flight will be conducted;
   (2) Navigation training at night in the vicinity of the airport where the solo flight will be conducted; and
   (3) An endorsement in the student's logbook for the specific make and model aircraft to be flown for night solo flight by an authorized instructor who gave the training within the 90-day period preceding the date of the flight.

Night solo flight is both airplane- and airport-specific, requiring previous instruction to be given in both areas. For an instructor pilot during an Alaskan winter, for example, the student could legally make solo cross-country flights only to airports previously flown to with an instructor and in the same type of aircraft. This regulation would suggest that night flying offers some increased danger to the inexperienced.

FAR 61.89 General Limitations [of Student Pilots]

(a) A student pilot may not act as pilot-in-command of an aircraft:
   (6) With a flight or surface visibility of less than 3 statute miles during daylight hours or 5 statute miles at night.

(7) When the flight cannot be made with visual reference to the surface; or

(8) In a manner contrary to any limitations placed in the pilot's logbook by an authorized instructor.

This one protects the fanny of the instructors. Students' solo privileges are based upon the instructor's certificate—if the student crashes, the instructor gets in trouble. It would be wise for instructors to view these weather minimums as just that—legal minimums. The instructor may place any additional limitations on their students that they see fit (like the insurance companies do), which become legally binding. My students do not often fly at night, and then only during conditions of severe clear and a bright moon—after complying with the regulatory requirements, of course.

The following section offers some insight into the preparations and training required of a student pilot for a cross-country flight. As you read these, put yourself into the shoes of a student pilot, as viewed from the standpoint of an instructor. Are your preparations sufficient that more experienced pilots would let someone dear to them fly along with you?

FAR 61.93 Solo Cross-Country Flight Requirements

(c) [Endorsements must]°

(A) Specify the make and model of aircraft to be flown;

(B) State that the student's preflight planning and preparation is correct and that the student is prepared to make the flight safely under the known conditions; and

(C) State that any limitations required by the student's authorized instructor are met.

(D) Limitations on authorized instructors to permit solo cross-country flights. An authorized instructor may not permit a student pilot to conduct a solo cross-country flight unless that instructor has:

(1) Determined that the student's cross-country planning is correct for the flight;

(2) Reviewed the current and forecast weather conditions and has determined that the flight can be completed under VFR;

(3) Determined that the student is proficient to conduct the flight safely;

(4) Determined that the student has the appropriate solo cross-country endorsement for the make and model of aircraft to be flown.

(E) Maneuvers and procedures for cross-country flight training....A student pilot who is receiving training for cross-country flight in a single-engine airplane must receive and log flight training in the following maneuvers and procedures:

(1) Use of aeronautical charts for VFR navigation using pilotage and dead reckoning with the aid of a magnetic compass;

(2) Use of aircraft performance charts pertaining to cross-country flight;

(3) Procurement and analysis of aeronautical weather reports and forecasts, including recognition of critical weather situations and estimating visibility while in flight;

(4) Emergency procedures;

(5) Traffic pattern procedures that include area departure, area arrival, entry into the traffic pattern, and approach;

(6) Procedures and operating practices for collision avoidance, wake turbulence precautions, and windshear avoidance;

(7) Recognition, avoidance, and operational restrictions of hazardous terrain features in the geographical area where the cross-country flight will be flown;

(8) Procedures for operating the instruments and equipment installed in the aircraft to be flown, including recognition and use of the proper operational procedures and indications;

(9) Use of radios for VFR navigation and two-way communications;

(10) Takeoff, approach, and landing procedures, including short-field, soft-field, and crosswind takeoffs, approaches, and landings;

(11) Climbs at best angle and best rate; and

(12) Control and maneuvering solely by reference to flight instruments, including straight and level flight, turns, descents, climbs, use of radio aids, and ATC directives.

Wow. The rule on this subject was recently expanded, and this is still for *student pilots.* As you look at each of the skills listed above, consider your own abilities in relation to them. Night operations specifically address the topics in lines (7) and (12).

FAR 61.101 Recreational Pilot Privileges and Limitations

...(d) Except as provided in paragraph (h) of this section, a recreational pilot may not act as pilot in command of an aircraft:

    (6) Between sunset and sunrise;

    (9) When the flight or surface visibility is less than 3 statute miles;

    (10) Without visual reference to the surface;

(h) For the purpose of obtaining additional certificates or ratings while under the supervision of an authorized instructor, a recreational pilot may fly as the sole occupant of an aircraft:

    (3) Between sunset and sunrise, provided the flight or surface visibility is at least 5 statute miles.

(i) In order to fly solo as provided in paragraph (h) of this section, the recreational pilot must meet the appropriate aeronautical knowledge and flight training requirements of section 61.87 for that aircraft. When operating an aircraft under the conditions specified in (h) of this section, the recreational pilot shall carry the logbook that has been endorsed for each flight by an authorized instructor who:

    (1) Has given the recreational pilot training in the make and model of aircraft in which the solo flight is to be made;

    (2) Has found that the recreational pilot has met the applicable requirements of section 61.87; and

    (3) Has found that the recreational pilot is competent to make solo flights in accordance with the logbook endorsement.

Recreational pilots are "students for life," operating under authorized supervision whenever desiring to exceed their limitations. Many pilots who operate airplanes for recreational purposes, even though not necessarily licensed as such, would be wise to consider their personal limitations as carefully.

FAR 61.107 Flight Proficiency [Private Pilots]

    (a) General. A person who applies for a private pilot certificate must receive and log ground and flight training from an authorized instructor on the areas of operation of this section that apply to the aircraft category and class rating sought.

    (b) Areas of operation.

        (1) For an airplane category rating with a single-engine class rating:

            (i) Preflight preparation;

            (ii) Preflight procedures;

            (ix) Basic instrument maneuvers;

            (x) Emergency operations;

            (xi) Night operations, except as provided in Sec. 61.110 of this part—

This section used to read:

    (a) In airplanes.

        (9) Night flying, including takeoffs, landings, and VFR navigation; and

        (10) Emergency operations, including simulated aircraft and equipment malfunctions.

"Emergency operations" include lighting and electrical failures, and the big one—engine failure. Instructors should be careful to teach a student how to identify and use a suitable landing area in the dark, emphasizing that the options there are limited, to say the least, and that some careful advance planning can be life-saving.

The current listing of FAR 61.107 (i–xi), pertaining to night flying and related skills, is duplicated for each category and class of aircraft rating sought, i.e., the same basic skills are required of a helicopter pilot as a multiengine pilot, where night flying is concerned. It is interesting that the old version of this section specifies some of the possible failures that may happen. Under the new section, failures like this are assumed under a blanket term "night flying operations." The point is, as far as FAA certification is concerned, you're expected to be able to cope with *anything*. The vagueness of the rule may prevent the FAA from being liable for any mishap you might have.

The following regulation has also changed, being recently broken into category and class designations as well as requiring new night and instrument experience. Here is the rule, as it used to read:

FAR 61.109 Airplane Rating: Aeronautical Experience

An applicant for a private pilot certificate with an airplane rating must have had at least a total of 40 hours of flight instruction and solo flight time which must include the following:

    (2) Three hours at night, including 10 takeoffs and landings for applicants seeking night flying privileges.

An applicant who does not meet the night flying requirement in paragraph (a)(2) of this section is issued a private pilot certificate bearing the limitation "Night flying prohibited." This limitation may be removed if the holder of the certificate shows that he has met the requirements of paragraph (a)(2) of this section.
It now reads:

FAR 61.109 Aeronautical Experience [Private Pilots]

(a) (1) Except as provided in section 61.110 of this part, 3 hours of night flight training in a single-engine airplane that includes—

    (i) One cross-country flight of over 100 nautical miles total distance; and

    (ii) 10 takeoffs and 10 landings to a full stop (with each landing involving a flight in the traffic pattern) at an airport.

(3) 3 hours of flight training in a single-engine airplane on the control and maneuvering of an airplane solely by reference to instruments, including straight and level flight, constant airspeed climbs and descents, turns to a heading, recovery from unusual flight attitudes, radio communications, and the use of navigation systems/facilities and radar services appropriate to instrument flight;

The rule continues with specifics and required experience in multiengine airplanes, helicopters, gyroplanes, powered lift, gliders, and so on, for those pilots preparing for a specific rating. Where an aircraft type operates in darkness, the FAA-required cross-country and instrument experience is essentially the same. Judging from the nature of the recent changes to this rule, the government feels that there is a corollary between night and instrument flight.

The 3 hours at night do not have to be all dual, although under most programs it works out that way. If the instructor is satisfied with the student's skills in the dark—satisfying legal requirements, it's okay for the student to fly solo at night. The ten landings are not required to be performed with the pilot as the sole manipulator of the controls—it is assumed that the pilot will initially need some help from the instructor while learning to land in the dark. Taking the regulation literally, however, it is possible for a private pilot applicant to be licensed for night flight without performing a landing in darkness, personally. FAR 61.57 specifies that the pilot be the sole manipulator of the controls during three recent night landings before carrying passengers, however, possibly requiring additional experience from such an ill-prepared private pilot.

FAR 61.110 makes an exception for pilots who learn to fly during the Alaskan summertime, when night conditions are not exactly available. The pilot certificate would bear restriction "night flying prohibited," until the experience requirements stated above are completed to the satisfaction of the FAA.

Moving on to commercial pilot certification, you will note that the differences in required night experience are rather slight.

FAR 61.125 Aeronautical Knowledge [Commercial Pilots]

(a) General. A person who applies for a commercial pilot certificate must receive and log ground training from an authorized instructor, or complete a home study course, on…

    (4) Meteorology to include recognition of critical weather situations, windshear recognition and avoidance, and the use of aeronautical weather reports and forecasts;

    (5) Safe and efficient operation of aircraft;

    (8) Significance and effects of exceeding aircraft performance limitations;

    (9) Use of aeronautical charts and a magnetic compass for pilotage and dead reckoning;

    (10) Use of air navigation facilities;

    (11) Aeronautical decision making and judgment;

    (12) Principles and functions of aircraft systems;

    (13) Maneuvers, procedures, and emergency operations appropriate to the aircraft;

    (14) Night and high-altitude operations;

    (15) Procedures for operating within the National Airspace System;

So far, the differences between commercial pilots and their private counterparts are mostly theoretical.

FAR 61.129 Aeronautical Experience

    (a) ...a person who applies for a commercial pilot certificate with an airplane category...rating must log at least...

        (3) 20 hours training...that includes at least—

            (i) 10 hours of instrument training...

        (4) 10 hours of solo flight...which includes at least—

            (ii) 5 hours in night VFR conditions with 10 takeoffs and 10 landings (with each landing involving a flight in the traffic pattern) at an airport with an operating control tower.

Suddenly, the budding commercial pilot is sent out to *fly solo in the darkness.* What better way to guarantee that the pilot is (as suggested in the previous version of this rule) "the sole manipulator of the controls." Of more interest, perhaps, because of its strong relationship to night flying, is the experience requirement for instrument training—over triple that required for private pilots. Looking further up on the skills and experience ladder, one would expect to see an instrument rating required of airline transport pilots, in addition to another plethora of night experience, which is not far off, as the ATP license is, in effect, a glorified instrument rating.

FAR 61.159 Aeronautical Experience: Airplane Category Rating

    (a) ...a person who is applying for an airline transport pilot certificate with an airplane category and class rating must have at least 1,500 hours of total time as a pilot that includes at least:

    (1) 500 hours of cross-country flight time.

    (2) 100 hours of night flight time.

    (3) 75 hours of instrument flight time, in actual or simulated instrument conditions...

    (4) 250 hours of flight time in an airplane as a pilot in command, or as second in command performing the duties and functions of a pilot in com-

mand while under the supervision of a pilot in command or any combination thereof, which includes at least—

(i) 100 hours of cross-country flight time; and

(ii) 25 hours of night flight time.

(5)(b) A person who has performed at least 20 night takeoffs and landings to a full stop may substitute each additional night takeoff and landing to a full stop for 1 hour of night flight time to satisfy the requirements of paragraph (a)(2) of this section; however, not more than 25 hours of night flight time may be credited in this manner.

Apparently, the FAA feels that a night landing carries as much challenge as spending an hour airborne in the dark, enabling a pilot to have only 50 hours total night experience and 45 landings to a full stop. Touch and goes don't count at night, unless the plane is brought to a halt on the runway. This probably stems from airplanes with conventional (taildragger) landing gear, which become more difficult to control as they slow down.

FAR 61.167 Privileges [Airline Transport Pilots]

(a) A person who holds an airline transport pilot certificate is entitled to the same privileges as those afforded a person who holds a commercial pilot certificate with an instrument rating....

Like I said, it's a glorified instrument rating.

So, whether professional or private, you've just read the minimum requirements demanded by the mighty FAA for certifying a pilot for night flight—really not that much different from going to college; you jump through enough hoops and pow!, you're certifiable. I suppose in that sense, we're all a little nuts. Having become certified, the pilot who actually attempts to *fly in the dark* faces an additional set of rules. For the sake of simplicity, we'll look primarily at FAR 91. There are, of course, additional regulations governing charter outfits and airlines, which require pilots to carry a manual with them that dictates rules ranging from flying at night, to how to cut their hair. For the sake of my own *certifiability,* we'll stick with part 91 until it begins discussing minimum equipment requirements, then stretch the comparison over to part 135.

FAR 91.3(a)

The pilot in command of an aircraft is directly responsible for, and is the final authority as to, the operation of that aircraft.

(b) In an in-flight emergency requiring immediate action, the pilot in command may deviate from any rule on this part to the extent required to meet that emergency.

This rule protects the fanny of the FAA. They don't want an airplane to crash because of some rule that they enforce, so pilots are given the authority to break rules as necessary to avoid in-flight difficulty. Unfortunately, the FAA often tries to second-guess the pilot who has violated rules in the aftermath of a crash.

Flying at night, a pilot may easily break VFR cloud avoidance rules by flying into clouds before the pilot realizes the clouds are there. According to this rule, the VFR pilot is legally square if steps are immediately taken to escape from the resulting IFR conditions.

FAR 91.7(b)

The pilot in command of a civil aircraft is responsible for determining whether that aircraft is in condition for safe flight. The pilot in command shall discontinue the flight when un-airworthy mechanical, electrical, or structural conditions occur.

An engine failure would constitute one of these, and, single-engine at night, the airplane may be more willing to abide by this rule than the pilot. Also, since external lighting is required at night, a night preflight must include a walk-around to inspect the condition and operation of the aircraft lights.

FAR 91.13(a)

No person may operate an aircraft in a careless or reckless manner so as to endanger the life or property of another.

 (b) No person may operate an aircraft, other than for the purpose of air navigation, on any part of the surface of an airport used by aircraft for air commerce (including areas used by those aircraft for receiving or discharging persons or cargo), in a careless or reckless manner so as to endanger the life or property of another.

I love this one. It says that careless or reckless flying is possibly *legal,* provided you own the plane, you're alone, and you don't endanger someone else's property on the ground. I wonder if the FAA would call some types of night operations "reckless."

FAR 91.103

Each pilot in command shall, before beginning a flight, become familiar with all available information concerning that flight. This information must include—

 (a) For a flight under IFR, or a flight not in the vicinity of an airport, weather reports and forecasts, fuel requirements, alternatives available if the planned flight cannot be completed, and any known traffic delays of which the pilot in command has been advised by ATC;

 (b) For any flight, runway lengths at airports of intended use, and the following takeoff and landing distance information;

  (1) …takeoff and landing distance data

  (2) …information appropriate to the aircraft, relating to aircraft performance under expected values of airport elevation and runway slope, aircraft gross weight, and wind and temperature.

Interestingly, this section mentions nothing specific about airport or runway lighting or other night-related equipment, except under the words "all available information concerning that flight."

FAR 91.111 (a)

No person may operate an aircraft so close to another aircraft as to create a collision hazard.

Night formation flight is legal—provided you can argue against the collision hazard if something goes wrong—otherwise you might be branded "reckless."

FAR 91.119

Except when necessary for takeoff or landing. No person may operate an aircraft below the following altitudes:

    (a) Anywhere. An altitude allowing, if a power unit fails, an emergency landing without undue hazard to persons or property on the surface.

    (b) Over congested areas. Over any congested area of a city, town, or settlement, or over any open air assembly of persons, an altitude of 1,000 feet above the highest obstacle within a horizontal radius of 2,000 feet of the aircraft.

    (c) Over other than congested areas. An altitude of 500 feet above the surface, except over open water or sparsely populated areas. In those cases, the aircraft may not be operated closer than 500 feet to any person, vessel, vehicle, or structure.

How high do you have to fly, at night, in order to satisfy the requirements of line (a)?

FAR 91.121(a)

Each person operating an aircraft shall maintain the cruising altitude or flight level of that aircraft, as the case may be, by reference to an altimeter that is set, when operating—

    (1) Below 18,000 feet MSL, to—

        (i) The current reported altimeter setting of a station along the route and within 100 nautical miles of the aircraft;

        (ii) If there is no station within the area prescribed in paragraph (a)(1)(i) of this section, the current reported altimeter setting of an appropriate available station; or

        (iii) In the case of an aircraft not equipped with a radio, the elevation of the departure airport or an appropriate altimeter setting available before departure; or

    (2) At or above 18,000 feet MSL, to 29.92" Hg.

It's possible to argue that your altimeter may be less accurate in the dark because local altimeter settings might not be available after hours.

FAR 91.123

(a) When an ATC clearance has been obtained, a pilot in command may not deviate from that clearance, except in an emergency, unless that pilot obtains an amended clearance. However, except in Class A airspace, this paragraph does not prohibit that pilot from canceling an IFR flight plan if the operation is being conducted in VFR weather conditions. When a pilot is uncertain of an ATC clearance, that pilot must immediately request clarification from ATC.

(b) Except in an emergency, no person may operate an aircraft contrary to an ATC instruction in an area in which air traffic control is exercised.

(c) Each pilot in command who, in an emergency, deviates from an ATC clearance or instruction shall notify ATC of that deviation as soon as possible.

This rule does not protect the pilot. Pilots are expected to correct erroneous clearances issued at the error of the controller. Vectors assigned over unknown terrain, at night, should demand the pilot's full attention. There have been cases where VFR traffic has been assigned a heading which, if left unchanged, would take the airplane directly into a mountain. Pilots are required to "speak up" in such circumstances.

FAR 91.125 ATC Light Signals

ATC light signals have the meaning shown in the following table.

Color and Type of Signal	Meaning with Respect to Aircraft on the Surface	Meaning with Respect to Aircraft in Flight
Steady green	Cleared for takeoff	Cleared to land
Flashing green	Cleared to taxi	Return for landing (to be followed by steady green at proper time)
Steady red	Stop	Give way to other aircraft and continue circling
Flashing red	Taxi clear of runway in use	Airport unsafe—do not land
Flashing white	Return to starting point on airport	Not applicable
Alternating red and green	Exercise extreme caution	Exercise extreme caution

If you've never seen or used an ATC light signal, ask the tower to show you one. You'll find them to be quite faint in daylight, and, although better at night, you have to be looking carefully to see them.

For the sake of clarification, we'll look at airspace classifications from a historical and tabular perspective. They'll be easier to remember that way.

The system got started when multiple airplanes began to converge with regularity on the same airports. It was any pilot's best guess as to which way the wind blew and any pilot's choice how to maneuver to the best position for landing. Many near misses and a few midairs later, somebody got the bright idea of a traffic pattern. Everybody looks at the windsock, and maneuvers to merge with the downwind leg of a traffic pattern; that puts the airplanes single-file in line for the runway, and since they all maintain spacing within gliding range of the runway, everyone is safe awaiting their turn to land (an emergency like an engine failure would give priority to the one in trouble). The traffic pattern seemed to appease a lot of anxious pilots who otherwise flew in a blistering hurry to land—engines of the time were not very reliable. A great idea, the traffic pattern, still in operation today. It's now defined with the requisite entries, exits, and the stipulation that all turns be to the left unless marked or indicated otherwise (91.126). This is the essence of operations at airports in Class G airspace (no tower, uncontrolled weather). Weather minimums became established at 1-mi visibility and clear of clouds (IFR, by most definitions).

Since poor weather occasionally interfered with the traffic pattern at busy airports, and given that IFR traffic did not necessarily follow traffic pattern procedure, certain airports needed a fix for when the weather was marginal. Initially called a "control zone" class E airspace, 91.127 stipulates VFR minimums of 3 mi and wider separation from clouds in order to facilitate "see and avoid" doctrine. Otherwise the pilot would have to use IFR procedure or special VFR procedures to find the runway in poor weather, both of which eliminated the pattern by simply allowing one airplane into the area at a time. Easily controlled from a distance by a controller who couldn't even see the area (like a flight service station), a "clearance" allowed a single airplane to fly into the area, locate and use the runway; then the pilot would notify the controller that the plane had landed safely, when a clearance could be issued to the next airplane. It worked in reverse for departing traffic. Things worked fine for quite a while this way, or at least on the bad weather days.

Anyone who flew in a traffic pattern on a beautiful (and busy) Saturday morning, however, soon found the normal traffic pattern procedure to be inadequate—the pattern can comfortably accommodate only so many airplanes before it becomes uncomfortably large. At busy airports, somebody got the bright idea to sit on top of a scaffold with a pair of binoculars, a radio, and a light signaling device (not many airplanes had radios). The runway and traffic pattern fell under the jurisdiction of this person in the makeshift tower. The controller had the option of manipulating the flow of traffic in the pattern and on the runway, in a discretionary manner, to facilitate the flow of airplanes to the airport. After a great deal of practice and trial and error, the tower system effectively permitted an airport to handle a much higher volume of traffic. The dimensions of this "airport traffic area" were dictated by the useful range of a good pair of binoculars—5 miles. Soon, a height designation followed—up to but not including 3000 ft—the controller did not want to be hassled with airplanes traversing the area at or above the VFR cruising altitudes. These dimensions have since changed (4.2 mi and 2500 ft), as has the name—class D airspace (91.129).

The controller served both as a director of traffic and as an extra pair of eyes on the ground. Light signals became unwieldy in the face of heavy traffic and soon gave way to the radio. Participation in the system suddenly required the airplanes to be equipped with radios, too—and for the pilots to take the initiative in establishing communication with the tower—assuming they wanted to use its runways. Thus we see the progression from simple, uncontrolled traffic patterns to dictated weather minimums, to the need for a tower and two-way radio communications.

Unfortunately, air traffic congestion continued to grow. Tower controllers became saturated, and the only viable solution was the opportunity to see traffic further out—extend the controllers' reach to a point well beyond the traffic pattern and its immediate vicinity. Radar soon surpassed the reach of binoculars, and an approach controller came into existence. Initially, pilots' participation in the system was voluntary—a pilot could ignore the approach controller and go right on in and talk to the tower. The approach controller provided a "service," using radar to assist pilots in avoiding other air traffic. With the burgeoning traffic, however, the system soon phased into mandatory participation of all air traffic. Radar controllers governed the areas reaching out from busy airports for 20 mi or more. Class C airspace is born (91.130).

Mandatory participation in the air traffic control service requires airplanes to be equipped with, of course, the requisite two-way radio and to establish communication with the controllers before entering the area. The current radar system also requires the use of a radar transponder—only lately requiring altitude encoding functions, as well. This is because the radar image is significantly filtered for the benefit of the controllers. Ground clutter, clouds, birds, and traffic on the freeway are filtered out, leaving only the air traffic visible on the screen. A computer-generated "target" appears where the radar suspects an aircraft to exist. The same target is drawn for a jumbo jet as for very small general aviation traffic. Airborne radar transponders return a signal that is codified in the ground radar computer to list the type, registration, destination, whether IFR, or VFR, groundspeed and altitude, among other notes, which the controller may then attach to the proper target. Target identification is simplified by the IDENT feature on most transponders. You hit that button at the controller's request, and your generic target is highlighted so that the controller may attach a list of notes to follow your airplane around on the screen. So, in a radar environment, a transponder is a must, along with the two-way communication.

Guess what? Things got busier still. The next option short of making landing reservations was to restrict entry into the radar-controlled airspace surrounding a busy airport. A pilot could IDENT and talk to the controllers all day, but until the controller says "cleared into the _____ airspace," the pilot is left to hold on the other side of the boundary. In this way, the controller may restrict the traffic flowing to the runway. This is the basis of Class B airspace: two-way communication, transponder, and a clearance to enter (91.131).

Thankfully, the final and most restrictive type of airspace does not exist around the busiest airports—not yet. For the moment, class A airspace exists only above 18,000 ft. To penetrate it, all the above equipment and procedures apply, with the addition of the requirement for an IFR flight plan—which is, in essence, a reservation. The IFR system allows air traffic to be scrutinized and manipulated from the moment of takeoff and followed all the way to landing. An airplane headed into the busiest airports in the country may be given delays and holds prior to takeoff or en route while still hundreds of miles away, asked to speed up, slow down, climb, turn, descend, or whatever would assist the controller in massaging the hundreds of converging airplanes into a line, single-file, for the runways. Class A airspace (91.135) is a long way from the traffic pattern, no?

FAR 91.137(a)

The Administrator will issue a Notice to Airmen (NOTAM) designating an area within which temporary flight restrictions apply and specifying the hazard or condition requiring their imposition, whenever he determines it is necessary in order to—

(1) Protect persons and property on the surface or in the air from a hazard associated with an incident on the surface;

(2) Provide a safe environment for the operation of disaster relief aircraft; or

(3) Prevent an unsafe congestion of sightseeing and other aircraft above an incident or event which may generate a high degree of public interest.

The Notice to Airmen will specify the hazard or condition that requires the imposition of temporary flight restrictions.

Line (3) is interesting here. Many sporting events, public gatherings, and fireworks happen at night. The airspace over a big stadium can be a congested mess of blimps, banner-towers and sightseers, and few actually receive temporary flight restrictions. In the end, in spite of all the FAA's noise and authority, the responsibility for air safety rests upon *your* shoulders.

FAR 91.151(a)

No person may begin a flight in an airplane under VFR conditions unless (considering wind and forecast weather conditions) there is enough fuel to fly to the first point of intended landing and, assuming normal cruising speed—

    (1) During the day, to fly after that for at least 30 minutes; or

    (2) At night, to fly after that for at least 45 minutes.

        (b) No person may begin a flight in a rotorcraft under VFR conditions unless (considering wind and forecast weather conditions) there is enough fuel to fly to the first point of intended landing and, assuming normal cruising speed, to fly after that for at least 20 minutes.

The FAA agrees with Chap. 2—you need more fuel to fly in the dark. There are fewer suitable runways and bigger unknowns. They figure that another 15 minutes of fuel should do the trick—if you were Lindbergh, would an extra 15 minutes be enough?

FAR 91.155 Basic VFR Weather Minimums

(a) Except as provided in paragraph (b) of this section and 91.157, no person may operate an aircraft under VFR when the flight visibility is less, or at a distance from clouds that is less, than that prescribed for the corresponding altitude and class of airspace in the following table:

Airspace	Flight visibility	Distance from clouds
Class A	Not Applicable	Not Applicable
Class B	3 statute miles	Clear of clouds
Class C	3 statute miles	500 feet below. 1000 feet above. 2000 feet horizontal.
Class D	3 statute miles	500 feet below. 1000 feet above. 2000 feet horizontal.
Class E: Less than 10,000 feet MSL	3 statute miles	500 feet below. 1000 feet above. 2000 feet horizontal
At or above 10,000 feet MSL	5 statute miles	1,000 feet below. 1000 feet above. 1 statute mile horizontal.
Class G: 1,200 feet or less above the surface (regardless of MSL altitude)		
Day, except as provided in § 91.155 (b)	1 statute mile	Clear of clouds.

Airspace	Flight visibility	Distance from clouds (*Continued*)
Night, except as provided in § 91.155 (b)	3 statute miles	500 feet below. 1000 feet above. 2000 feet horizontal.
More than 1,200 feet above the surface but less than 10,000 feet MSL		
Day	1 statute mile	500 feet below. 1000 feet above. 2000 feet horizontal.
Night	3 statute miles	500 feet below. 1000 feet above. 2000 feet horizontal.
More than 1,200 feet above the surface and at or above 10,000 feet MSL	5 statute miles	1000 feet above. 1000 feet below 1 statute mile horizontal.

(b) Class G airspace. Notwithstanding the provision of paragraph (a) of this section, the following operations may be conducted in Class G airspace below 1,200 feet above the surface:

    (1) Helicopter. A helicopter may be operated clear of clouds if operated at a speed that allows the pilot adequate opportunity to see any air traffic or obstruction in time to avoid a collision.

    (2) Airplane. When the visibility is less than 3 statute miles but not less than 1 statute mile during night hours, an airplane may be operated clear of clouds if operated in an airport traffic pattern within one-half mile of the runway.

(c) Except as provided in 91.157, no person may operate an aircraft beneath the ceiling under VFR within the lateral boundaries of controlled airspace designated to the surface for an airport when the ceiling is less than 1,000 feet.

(d) Except as provided in 91.157 of this part, no person may take off or land an aircraft, or enter the traffic pattern of an airport, under VFR, within the lateral boundaries of the surface areas of Class B, Class C, Class D, or Class E airspace designated for an airport—

    (1) Unless ground visibility at that airport is at least 3 statute miles; or

    (2) If ground visibility is not reported at that airport, unless flight visibility during landing or takeoff, or while operating in the traffic pattern is at least 3 statute miles.

(e) For the purpose of this section, an aircraft operating at the base altitude of a Class E airspace area is considered to be within the airspace directly below that area.

The foregoing is a mess, but exists this way in an effort to provide the least intrusive regulation. You can bet that a simplification of the system would be to require all air traffic to file IFR—eliminating all cloud and visibility requirements. So, the VFR weather minimums are really complicated to maintain your personal freedom—or at least some semblance of it. We have no choice but to memorize the minimums and see that we do not admit to violating them. In truth, the only sure way the FAA will know you breached one of these minimums is if you admit doing so or land below minimums at a runway that has recording weather-detection equipment.

FAR 91.157 Special VFR Weather Minimums

(a) Except as provided in appendix D, section 3, of this part, special VFR operations may be conducted under the weather minimums and requirements of this section, instead of those contained in 91.155, below 10,000 feet MSL within the airspace contained by the upward extension of the lateral boundaries of the controlled airspace designated to the surface for an airport.

(b) Special VFR operations may only be conducted—

(1) With an ATC clearance;

(2) Clear of clouds;

(3) Except for helicopters, when flight visibility is at least 1 statute mile; and

(4) Except for helicopters, between sunrise and sunset (or in Alaska, when the sun is 6 degrees or more above the horizon) unless—

(i) The person being granted the ATC clearance meets the applicable requirements for instrument flight under part 61 of this chapter;

(ii) The aircraft is equipped as required in Part 91.205 (d).

(c) No person may take off or land an aircraft (other than a helicopter) under special VFR—

(1) Unless ground visibility is at least 1 statute mile; or

(2) If ground visibility is not reported, unless flight visibility is at least 1 statute mile.

You and your plane need to be instrument-rated to accept a special VFR clearance at night.

FAR 91.167 Fuel Requirements for Flight in IFR Conditions

(a) Except as provided in paragraph (b) of this section, no person may operate a civil aircraft in IFR conditions unless it carries enough fuel (considering weather reports and forecasts and weather conditions) to—

(1) Complete the flight to the first airport of intended landing;

(2) Fly from that airport to the alternate airport; and

(3) Fly after that for 45 minutes at normal cruising speed or, for helicopters, fly after that for 30 minutes at normal cruising speed.

(b) Paragraph (a)(2) of this section does not apply if—

(1) Part 97 of this chapter prescribes a standard instrument approach procedure for the first airport of intended landing; and

(2) For at least 1 hour before and 1 hour after the estimated time of arrival at the airport, the weather reports or forecasts or any combination of them indicate—

(i) The ceiling will be at least 2,000 feet above the airport elevation; and

(ii) Visibility will be at least 3 statute miles.

This is interesting in its similarity to the night fuel reserve. The two are identical, except for the IFR requirement of an alternate.

FAR 91.173 ATC Clearance and Flight Plan Required

No person may operate an aircraft in controlled airspace under IFR unless that person has—

(a) Filed an IFR flight plan; and

(b) Received an appropriate ATC clearance.

Blundering into IFR conditions at night may require some quick radio work on the part of the pilot to establish legalities and join the IFR system. If you're unwilling to declare an emergency, you should not admit to being in IFR conditions until the clearance has been received. Otherwise, you could be held in violation. If in doubt, declare the unexpected IFR conditions, get the clearance, and fill out a NTSB/NASA accident/incident reporting form.

FAR 91.177 Minimum Altitudes for IFR Operations

(a) Operation of aircraft at minimum altitudes. Except when necessary for take-off or landing, no person may operate an aircraft under IFR below—

(2) (i) In the case of operations over an area designated as a mountainous area in part 95, an altitude of 2,000 feet above the highest obstacle within a horizontal distance of 4 nautical miles from the course to be flown; or

(ii) In any other case, an altitude of 1,000 feet above the highest obstacle within a horizontal distance of 4 nautical miles from the course to be flown.

This rule compensates for the inaccuracies of instrument navigation. If you plan to rely on this type of navigation while flying VFR, it might be wise to adapt your flight plan such that it maintains similar altitude margins, thus preparing yourself for unexpected IFR conditions.

SUBPART C—EQUIPMENT, INSTRUMENT, AND CERTIFICATE REQUIREMENTS

FAR 91.205 Powered Civil Aircraft with Standard Category U.S. Airworthiness Certificates: Instrument and Equipment Requirements

(a) General. Except as provided in paragraphs (c)(3) and (e) of this section, no person may operate a powered civil aircraft with a standard category U.S. airworthiness certificate in any operation described in paragraphs (b) through (f) of this section unless that aircraft contains the instruments and equipment specified in those paragraphs (or FAA-approved equivalents) for that type of operation, and those instruments and items of equipment are in operable condition.

(b) Visual-flight rules (day). For VFR flight during the day, the following instruments and equipment are required (*Fig. 3-1*):

(1) Airspeed indicator.

(2) Altimeter.

(3) Magnetic direction indicator.

(4) Tachometer for each engine.

(5) Oil pressure gauge for each engine using pressure system.

(6) Temperature gauge for each liquid-cooled engine.

(7) Oil temperature gauge for each air-cooled engine.

(8) Manifold pressure gauge for each altitude engine.

(9) Fuel gauge indicating the quantity of fuel in each tank.

(10) Landing gear position indicator, if the aircraft has a retractable landing gear.

(11) If the aircraft is operated for hire over water and beyond power-off gliding distance from shore, approved floatation gear readily available to each occupant and at least one pyrotechnic signaling device.

(12) An approved safety belt with an approved metal-to-metal latching device for each occupant 2 years of age or older.

(13) For small civil airplanes manufactured after July 18, 1978, an approved shoulder harness for each front seat.

(14) An emergency locator transmitter, if required by 91.207

(c) Visual flight rules (night) (*Fig. 3-2*). For VFR flight at night, the following instruments and equipment are required:

(1) Instruments and equipment specified in paragraph (b) of this section.

(2) Approved position lights.

(3) An approved aviation red or aviation white anti-collision light system on all U.S.-registered civil aircraft....In the event of a failure of any light of the anti-collision light system, oper-

Required equipment for day VFR, according to FAR 91

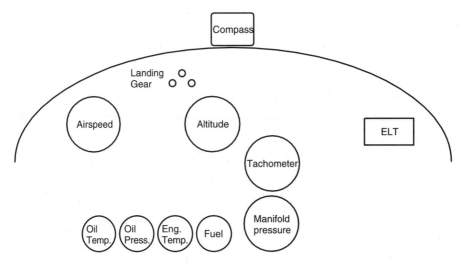

Fig. 3-1. *This panel is surprisingly basic and could be more so if the airplane had fixed gear and a fixed-pitch propeller. [See FAR 91.205 (b), (8) and (10)]*

Required equipment for night VFR, according to FAR 91

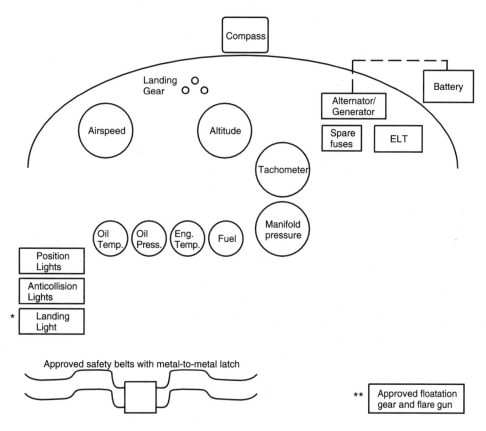

*If operated for hire.
**If operated over water, beyond power-off gliding distance from shore.

Fig. 3-2. *Legal night flight requires the presence of lighting and electrical equipment.*

ations with the aircraft may be continued to a stop where repairs or replacement can be made.

(4) If the aircraft is operated for hire, one electric landing light.

(5) An adequate source of electrical energy for all installed electrical and radio equipment.

(6) One spare set of fuses, or three spare fuses of each kind required, that are accessible to the pilot in flight.

(d) Instrument flight rules. For IFR flight, the following instruments and equipment are required (*Fig. 3-3*):

(1) Instruments and equipment specified in paragraph (b) of this section, and, for night flight, instruments and equipment specified in paragraph (c) of this section.

Required equipment for basic IFR, according to FAR 91

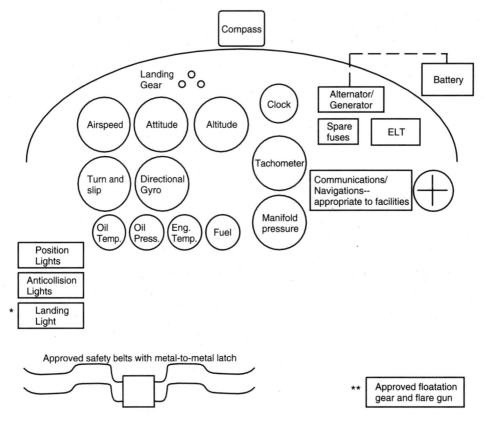

Fig. 3-3. *Instrument flight requires gyros and navigational equipment. Night flight could often qualify under IFR.*

(2) Two-way radio communications system and navigational equipment appropriate to the ground facilities to be used.

(3) Gyroscopic rate-of-turn indicator, except on the following aircraft:

(i) Airplanes with a third attitude instrument system usable through flight attitudes of 360 degrees of pitch and roll and installed in accordance with the instrument requirements prescribed in 121.305(j) of this chapter; and

(ii) Rotorcraft with a third attitude instrument system usable through flight attitudes of ±80 degrees of pitch and ±120 degrees of roll aid installed in accordance with 29.1303(g) of this chapter.

(4) Slip-skid indicator.

(5) Sensitive altimeter adjustable for barometric pressure.

    (6) A clock displaying hours, minutes, and seconds with a sweep-second pointer or digital presentation.

    (7) Generator or alternator of adequate capacity.

    (8) Gyroscopic pitch and bank indicator (artificial horizon).

    (9) Gyroscopic direction indicator (directional gyro or equivalent).

FAR 91.209 Aircraft Lights

No person may, during the period from sunset to sunrise (or, in Alaska, during the period a prominent unlighted object cannot be seen from a distance of 3 statue miles or the sun is more than 6 degrees below the horizon)—

    (a) Operate an aircraft unless it has lighted position lights (*Fig. 3-4a*);

    (b) Park or move an aircraft in, or in dangerous proximity to, a night flight operations area of an airport unless the aircraft—

        (1) Is clearly illuminated (*Fig. 3-4b*);

        (2) Has lighted position lights; or

        (3) Is in an area which is marked by obstruction lights;

    (c) Anchor an aircraft unless the aircraft—

        (1) Has lighted anchor lights (*Fig. 3-4c*); or

        (2) Is in an area where anchor lights are not required on vessels; or

    (d) Operate an aircraft, required by 91.205(c)(3) to be equipped with an anti-collision light system, unless it has approved and lighted aviation red or aviation white anti-collision lights (*Fig. 3-4d*). However, the anti-collision lights need not be lighted when the pilot in command determines that, because of operating conditions, it would be in the interest of safety to turn the lights off.

Fig. 3-4. *In an effort to improve dispatch reliability (where a broken bulb could cancel a flight), many airliners are equipped with two of each required navigation light. General aviation aircraft have simpler systems.*

Fig. 3-4b. (*Continued*)

Fig. 3-4c. (*Continued*)

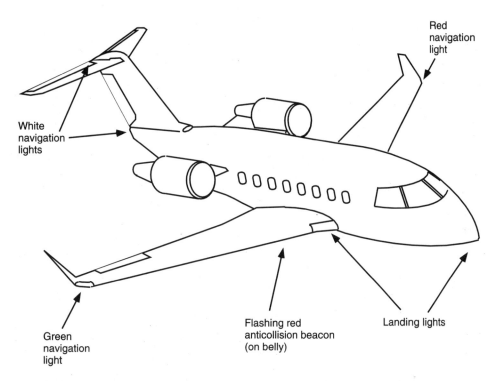

White
navigation
lights

Red
navigation
light

Green
navigation
light

Flashing red
anticollision beacon
(on belly)

Landing lights

Fig. 3-4 d. (*Continued*)

FAR 91.211 Supplemental Oxygen

(a) General. No person may operate a civil aircraft of U.S. registry—

(1) At cabin pressure altitudes above 12,500 feet (MSL) up to and including 14,000 feet (MSL) unless the required minimum flight crew is provided with and uses supplemental oxygen for that part of the flight at
those altitudes that is of more than 30 minutes duration;

(2) At cabin pressure altitudes above 14,000 feet (MSL) unless the required
minimum flight crew is provided with and uses supplemental oxygen
during the entire flight time at those altitudes; and

(3) At cabin pressure altitudes above 15,000 feet (MSL) unless each occupant of the aircraft is provided with supplemental oxygen.

This is included because the nighttime recommendation is that oxygen be used by pilots
at altitudes exceeding 5000 ft MSL—quite different from the daylight specifications.

FAR 91.213 Inoperative Instruments and Equipment

(a) Except as provided in paragraph (d) of this section, no person may take off
an aircraft with inoperative instruments or equipment installed unless the
following conditions are met:

(1) An approved Minimum Equipment List exists for that aircraft.

(2) The aircraft has within it a letter of authorization, issued by the FAA
Flight Standards district office having jurisdiction over the area in which

the operator is located, authorizing operation of the aircraft under the Minimum Equipment List. The letter of authorization may be obtained by written request of the airworthiness certificate holder. The Minimum Equipment List and the letter of authorization constitute a supplemental type certificate for the aircraft.

(d) Except for operations conducted in accordance with paragraph (a) or (c) of this section, a person may takeoff an aircraft in operations conducted under this part with inoperative instruments and equipment without an approved Minimum Equipment List provided—

　(1) The flight operation is conducted in a—

　　(i) Rotorcraft, nonturbine-powered airplane, glider, or lighter-than-air aircraft for which a master Minimum Equipment List has not been developed; or

　　(ii) Small rotorcraft, nonturbine-powered small airplane, glider, or lighter-than-air aircraft for which a Master Minimum Equipment List has been developed; and

　(2) The inoperative instruments and equipment are not—

　　(i) Part of the VFR-day type certification instruments and equipment prescribed in the applicable airworthiness regulations under which the aircraft was type certificated;

SUBPART D—SPECIAL FLIGHT OPERATIONS

91.303 Aerobatic flight—

No person may operate an aircraft in aerobatic flight—

(a) Over any congested area of a city, town, or settlement;

(b) Over an open air assembly of persons;

(c) Within the lateral boundaries of the surface areas of Class B, Class C, Class D, or Class E airspace designated for an airport;

(d) Within 4 nautical miles of the center line of any Federal airway;

(e) Below an altitude of 1,500 feet above the surface; or

(f) When flight visibility is less than 3 statute miles.

For the purposes of this section, aerobatic flight means an intentional maneuver involving an abrupt change in an aircraft's attitude, an abnormal attitude, or abnormal acceleration, not necessary for normal flight.

The interesting feature of this rule is what it does not contain—night aerobatics are legal.

FAR 91.305 Flight Test Areas—

No person may flight test an aircraft except over open water, or sparsely populated areas, having light air traffic.

Here too, conspicuous by its absence, is that night flight test is legal.

FAR 91.309 Towing: Gliders—

Also, legal at night.

FAR 91.311 Towing: Other than under 91.309—

Also, legal at night.

FAR 91.313 Restricted Category Civil Aircraft: Operating Limitations—
Crop dusting, firefighting, weather research flights, and so on, are *legal in the dark.*

FAR 135, SUBPART C—AIRCRAFT AND EQUIPMENT

FAR 135.159
No person may operate an aircraft carrying passengers under VFR at night or under VFR over-the-top, unless it is equipped with—
  (a) A gyroscopic rate-of-turn indicator…
  (b) A slip skid indicator.
  (c) A gyroscopic bank-and-pitch indicator.
  (d) A gyroscopic direction indicator.
  (e) A generator or generators able to supply all probable combinations of continuous in-flight electrical loads for required equipment and for recharging the battery.
  (f) For night flights—
    (1) An anti-collision light system;
    (2) Instrument lights to make all instruments, switches, and gauges easily readable, the direct rays of which are shielded from the pilots eyes; and
    (3) A flashlight having at least two size "D" cells or equivalent.
  (g) For the purpose of paragraph (e) of this section, a continuous in-flight electrical load includes one that draws current continuously during flight, such as radio equipment and electrically driven instruments and lights, but does not include occasional intermittent loads.

FAR 135.161 Radio and Navigational Equipment: Carrying Passengers under VFR at Night or under VFR Over-the-top.
  (a) No person may operate an aircraft carrying passengers under VFR at night, or under VFR over-the-top, unless it has two-way radio communications equipment able, at least in flight, to transmit to, and receive from, ground facilities 25 miles away.
  (c) No person may operate an aircraft carrying passengers under VFR at night unless it has radio navigational equipment able to receive radio signals from the ground facilities to be used.

FAR 135.163 Equipment Requirements: Aircraft Carrying Passengers under IFR
No person may operate an aircraft under IFR, carrying passengers, unless it has—
  (a) A vertical speed indicator;
  (b) A free-air temperature indicator;
  (c) A heated pitot tube for each airspeed indicator;
  (d) A power failure warning device or vacuum indicator to show the power available for gyroscopic instruments from each power source;
  (e) An alternate source of static pressure for the altimeter and the airspeed and vertical speed indicators;

(f) For a single-engine aircraft, a generator or generators able to supply all probable combinations of continuous in-flight electrical loads for required equipment and for recharging the battery;

(g) For multi-engine aircraft, at least two generators, each of which is on a separate engine, of which any combination of one-half of the total number are rated sufficiently to supply the electrical loads of all required instruments and equipment necessary for safe emergency operation of the aircraft.

(h) Two independent sources of energy (with means of selecting either), of which at least one is an engine-driven pump or generator, each of which is able to drive all gyroscopic instruments and installed so that failure of one instrument or source does not interfere with the energy supply of the remaining instruments or the other energy source.

[Note: Required equipment for night IFR includes a flashlight (*Fig. 3-5*).]

FAR 135.181

(a) Except as provided in paragraphs (b) and (c) of this section, no person may—

(1) Operate a single-engine aircraft carrying passengers over-the-top or in IFR conditions; or

(2) Operate a multi-engine aircraft carrying passengers over-the-top or in IFR conditions at a weight that will not allow it to climb, with the critical engine inoperative, at least 50 feet a minute when operating at the MEAs of the route to be flown or 5,000 feet MSL, whichever is higher.

(d) Without regard to paragraph (a) of this section, a person may operate an aircraft over-the-top under conditions allowing—

(1) For multi-engine aircraft, descent or continuance of the flight under VFR if its critical engine fails;

or

(2) For single-engine aircraft, descent under VFR if its engine fails.

The foregoing is the government's wish list. In addition, charter and scheduled operators must meet performance requirements that suit the route to be flown. These make sense. What's the use of a multiengine airplane that can't maintain a safe altitude on one engine?

Part 135 also stipulates rest requirements for the flight crew, in effort to prevent fatigue-induced cockpit errors. These are stipulated in Part 135, subpart F, and are somewhat beyond the scope of this chapter, but feel free to look them up on your own.

TECHNIQUE

Flying in accordance to FAA regulation is becoming increasingly difficult. Rules and regulations on the scheduled and charter carriers are so complicated that pilots may be tempted to moonlight as lawyers, giving a legal opinion for a buck. Regarding the methods used to stay legal, the only really valuable technique is to know the rules well, and if in doubt, don't admit to *anything*. For the sake of knowing the rules, here is the foregoing, restated in a nutshell, as they apply to basic night flight—avoiding the complexities of flight for hire.

Required equipment for night/IFR, according to FAR 135

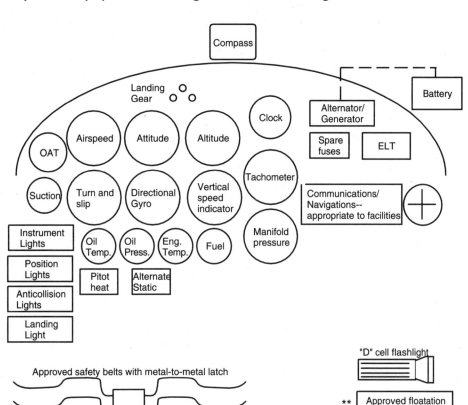

**If operated over water, beyond power-off gliding distance from shore.

Fig. 3-5. *Charter operators are legally required to bring a flashlight.*

Minimum Additional Night Equipment

- Approved position lights
- Approved anticollision lights
- Adequate source of electrical energy to power lights and radios
- A set of spare fuses

Minimum Additional Night Flight Preparation. The airplane must carry an additional 15 minutes of fuel over daylight reserves, for a total of 45 minutes at normal cruising speed.

Night Pilot Currency Requirements. Pilots must have logged three takeoffs and three landings to a full stop, at night, in the same category and class of airplane to be used—unless their licenses say "Night flying prohibited," in which case the logging of night landings could be taken as an admission of guilt.

Night Weather Minimums—Where They Differ from Daylight. All the visibility minimums that were 1 statute mi in daylight, become 3 statute mi at night. Anything that is "clear of clouds" in daylight, becomes the standard: 1000 ft above, 500 ft below and 2000 ft horizontal separation. Special VFR requires IFR equipment and an appropriately rated pilot.

Summary. That's it, unless you're trying to jump through the hoops legally necessary to sell a ticket on your plane. As far as the FAA is concerned, you need lights, a little experience, and a little more fuel, and please stay out of the weather. From what we have discussed already, this is only a legal minimum—the real needs of safety dictate a bit more.

FURTHER READING

Federal Aviation Regulations, chapters 61, 91, 135, 121
Airman's Information Manual

4
Physiology at night

Christmas eve comes quietly, as if everyone is holding their breath for the big day. Southern California sparkled with decorative lights, closed businesses and the radio playing carols for an occasional long-running party. It was four in the morning, December 24th. Most people rested quietly in their homes. Whether under a blanket of snow or Southwestern warmth, the holidays have an effect on everybody; and for Brett it meant flying almost nonstop for the last 3 days.

The little air freight company won, for the first time, a lucrative UPS feeder contract, for the mountain of packages sent during the holiday season. The pilots at first felt like Santa Claus, shuttling Yuletide cargo hither and yon, enjoying the opportunity to fly and earn money. The opportunity more than matched the size of the company, however, with far more loot to carry than they had planes and pilots for. To win future contracts, the company *had* to satisfy the demand, so the boss made a few demands of the chief pilot. The chief pilot made a few demands of the freight pilots, and Brett had to fly. No fussing about FAA mandated rest periods; no slacking on the job. "The plane has to be in Bakersfield in the morning. *There is no one else available to fly it, so we're counting on you to get it done…Got it?*"

So Brett had to fly. He'd flown all night, every night, and all day for the past 3 days. Sleeping at outstations, between flights, for only a few hours at a time, eating on the plane and loading freight in between. He didn't feel like Santa Clause at all; he was so tired that he couldn't feel anything. Christmas eve, Silent Night—holiday spirit everywhere, and all he cared was that it would soon be over. He didn't want to fly, shouldn't fly, felt way too tired, had been awake and working almost 40 hours, with pitifully small amounts of sleep in between shifts—no more than a couple of hours at a time. He trudged off to the Piper Navajo, waiting quietly in the darkness of the Ontario airport ramp. His limbs felt heavy, making the walking seem difficult. It was not particularly hard for him to stay awake while loading the plane—the physical exertion and cool air helped—but he felt dangerously fatigued. He worked under the floodlights that brightened the ramp around the plane, but night encroached upon the shadows, like dark dreams waiting to jump a tired passer-by. *Dangerous to be flying like this—rather sleep, could pass out right here, on a crate.* He settled heavily into the familiar cockpit, glad to be through with loading the plane—glad it would soon be over. A flight to Bakersfield, and it's Christmas eve, the last day of the contract—heck, it's Christmas eve now, already four-thirty in the morning—get there and the work is done—rest all you want.

It's dark outside, pitch black, early morning dark, and the weather reports indicate low clouds over the route with dense ground fog at Bakersfield—the usual for this time of year. Weather like this makes a pilot earn his keep—approaches to minimums—if the controllers will cooperate and call it at minimums—they'd have to stretch it a bit. A challenge at any time, but with a sleepy pilot, utterly dangerous. Brett launched into the low overcast, instantly comfortable with the throbbing of the two engines outside. The normal mechanical vibrations relaxed him, soon breaking out of the weather, flying above the clouds. The flying soothed him; it felt peaceful, restful…nice. "CALAIR_____, CALAIR_____, SO-CAL APPROACH, LISTEN UP, WAKE UP! TURN RIGHT, HEADING 285, HEADING 285—ACKNOWLEDGE!" The controller shouted pretty loud, practically screaming in the radio—and Brett woke up. *Shouldn't have been sleeping.* He felt momentarily confused at his surroundings—a little surprised to be in a cockpit, and IFR, at that. He tried to concentrate on the task—*right turn, heading 285. Not so hard. Tired.* Brett's voice was slurred in his communications. "Ah, roger—right heading 285." The controller continued to yell on the radio. *Does he have to be so loud? How did the heading get to 240?* He turned back and overshot, rolling out at 300, or so. *Heading 285, ah—285. Wow, tired—feel almost drunk.*

Brett squinted and rubbed his eyes. Sitting comfortably in the cockpit, he didn't have to move much. His body felt like a sandbag, weighing heavily, painfully, in the seat. It ached. The fatigue he felt was pain; a dull, almost overwhelming ache that spread outward from his bones—his body resisted every movement. His eyes felt the heaviest of all. Intellectually, he knew that to fall asleep meant certain death; physically, his eyelids weighed a ton each—impossible to keep open. He rubbed them hard, the pressure of hands against his eyeballs making him see lights. He jerked awake suddenly, kicking one of the rudder pedals and feeling out of balance. *Have to stay awake! Keep my eyes open— don't sleep!* He slapped his knees and adjusted himself in the cockpit—and jerked awake again, almost in the middle of it all. He was falling asleep several times a minute in the process of trying to stay awake, fighting a losing battle.

When he next looked out the windows, he was shocked to find the plane between layers, flying in a 90-degree bank to the right! He flinched on the controls. *Roll out, roll out—reference the attitude indicator and roll out—but the attitude gyro shows wings level. Back outside—definite 90-degree bank; attitude indicator still showing level.* Brett experienced a momentary jolt of adrenaline as he recognized the trouble. *Never had vertigo backward before. Always looks level outside, when the indicator is in a bank. This feels weird. Focus on the attitude indicator. Concentrate on it until it fills your vision. Crawl inside that thing until what it shows, feels like level.* "CALAIR_____, CALAIR_____, SO-CAL APPROACH, WHERE ARE YOU GOING?! MAINTAIN HEADING 290, NOW!" The controller spoke loudly, urgently, knowing that the Calair pilot was in serious trouble. He tried to help, as Brett struggled to fly the plane. *Why is it so hard to keep a heading?*

As the pressure of the flight continued to build, his thoughts fragmented; he had difficulty thinking at all. *Never been so tired. Have no business flying an airplane. So fatigued—my body hurts! Rather sleep than fly. Rather sleep than do anything—such a bother to stay awake. The pesky controller is annoying—won't let me sleep. Don't want to fly the plane…just want to close my eyes….*

Brett was barely conscious, not quite rational, and losing the battle. The plane arrived over Bakersfield at sunrise, the bright morning light glaring in the fog, which caused the visibility to go down. The tower generously called the visibility at minimums so Brett could attempt the approach. There was a good chance of a missed approach, with the alternate about an hour's flight away—and Brett was about gone. Flying erratic this way, IFR—heavy IFR—*hard to think, it's crazy.* He could barely hold a heading, couldn't help drifting off, almost constantly nodding off—he was in agony; the fatigue of long days focused in his eyelids. He could not keep them open much longer—and stood a good chance of crashing if he flew another hour. He had to land on the first attempt, or…

Somewhere in his sleep-crazed mind, Brett came to grips with the situation. *I make this approach, or crash—so, in a few minutes I'll get to close my eyes, one way or another…I'll get to close my eyes.* He didn't really care how the flight turned out, from then on.

Turning onto the localizer, he caught the course and managed, erratically, to fly the approach. The controllers were indeed generous with their estimate of the weather—*nice of them to stretch the minimums like that.* Brett found the runway with his tires and rolled to a stop on the ramp. He opened the cockpit door and felt a cool breeze on his face.

He could finally sleep….

BACKGROUND INFORMATION

Brett survived a close brush with fatigue that could easily have been fatal. The most remarkable thing for him, about that experience, was that he arrived at the point where he did not care if he crashed and died, or not—either way, it meant that he could get some rest. It was fatigue-clouded judgment that made an almost suicidal attitude like that possible. Your body has a basic need for sleep—like it needs to eat and breathe. Going long enough without sleep can result in hallucinations, unwilling lapses into sleep, poor motor control, impaired judgment, severe mood swings, and mental psychosis—all of which are bad for flying. But sleep and fatigue are only part of the issue; darkness also affects your eyes (obviously), hearing, balance, speech, emotions, metabolism and decision-making ability, to name a few.

Circadian Rhythms and Metabolism

From the beating of your heart, to the gentle breathing of your lungs, to the timing of your meals, to the time you awake in the morning; your body demonstrates a remarkable sense of rhythm. These rhythms are important and difficult to ignore. Take your breathing, for example. You breathe in and out constantly throughout the day. When you think about it, you can upset the rhythm by holding your breath, but your body will complain strongly and eventually lapse into unconsciousness, where it will resume normal breathing again. You can breathe faster than normal if you wish, but unless there is a need, such as to compensate for exercise, you will soon get light-headed, and eventually lapse into unconsciousness again, where your body will resume breathing normally, keeping the rhythm. Your heart beats at an even rhythm and we all know what happens if *that* stops.

There is a much slower rhythm to your sleeping. Your body would like to sleep a few hours every day—maintain a circadian rhythm. *Circadian* means "about a day," referring to the daily cycles of your body, and it refers to more than just sleeping. Your bowels, for example, maintain a rhythm of their own—you have a little influence in deciding where to empty them, but *you can't hold out forever.* Sleep is similar to this. You may avoid sleeping altogether, if you wish, but the consequences are similar to restraining your bladder. Eventually, your control of the situation will fade, and your body will sleep anyway. In the process of becoming incapacitated by lack of sleep, you may experience all sorts of symptoms, many of which border on the psychotic.

Your metabolism functions in concert with the rest of your body. *Metabolism* refers to your body's rate of energy consumption—a throttle setting, if you will. A generally high metabolic rate could be demonstrated by a hyperactive child, and a low rate is easily evidenced by a couch potato. Your metabolic rate will change with response to your body's needs. It slows down when you sleep (Fig. 4-1). If you are awake at night, out of rhythm with what your body expects you to do, your metabolic rate may still slow down. This could make you feel more tired than normal, lacking the energy you'd expect in the daytime. People who skip a night's sleep feel tired at night but then seem to feel some improvement in the morning, because their metabolism is picking up for the daylight hours, as it is accustomed to doing. Because of this, you may assume that your state of alertness will generally be lessened during the dark hours, unless you take pains to change your body's rhythms.

Your body's rhythms are not set in concrete. They can be altered by modifying your habits. You automatically breathe faster when you exercise. Eat bad food, and your bowels may function at warp speed. If you get sick, your body may need to sleep much more than normal. Sleep a lot during the day, and you may stay awake for much of the night. With discipline, you might reverse the normal day-night cycles altogether. Farmers have, for years, awakened at very early hours and retired to bed long before nightfall. Night cargo pilots train themselves to sleep in the day, awaking a bit before sundown, enabling them to keep the same hours as a bat. These things do not come without consequences,

Fig. 4-1. *As the day draws to a close, your metabolism slows down.*

however. Brett, for example, quit Calair a few months after his hair-raising flight. He was *tired,* for lack of a better word, of working nights. After adopting a daylight schedule, he needed almost 2 months to fully recover and feel "normal" again. He noticed the reversal of several changes that occurred as he flew the night freight, unawares. The dark circles under his eyes disappeared. His personality changed—he became more vibrant, energetic. He felt healthier. All these things were different when he worked at night; he did not notice them until he had stopped.

Many drugs, to the delight of rock stars and Hollywood, are available to alter sleeping rhythms artificially. Narcotics, alcohol, allergy medicines, weight-control pills, cold remedies, and literally thousands of others can all affect your ability to sleep or stay awake, changing your level of alertness and often clouding your judgment. Some of the popular illegal drugs may also affect a pilot's attitude, causing inappropriate responses to dire situations. For example, some drugged pilots being tested as they fly a simulator actually *laughed* as they impacted the ground. If a pilot expects to maintain an adequate state of alertness for a safe flight, by all means, seek a professional opinion on the viability of questionable drugs in the cockpit.

Fatigue

A pilot who looks or feels generally tired, slow of thought, withdrawn, and unresponsive is probably suffering some of the effects of fatigue. Fatigue, in itself, has proved difficult to describe and even harder to quantify. NTSB investigations often list pilot fatigue as a contributing factor in accidents, and never as a cause, because of the difficulty of measuring or even proving the existence of the condition. An investigator can only assume that fatigue played a part in a crash because of circumstantial evidence: The pilot did not sleep well for days before, was subjected to long hours at work, and appeared to show symptoms of a cold—all factors that could cumulatively lead to performance degradation in the cockpit as a result of fatigue. Some of the causes of fatigue include, but are not limited to:

- Disruption of circadian patterns
- External stresses unrelated to flying
- Inadequate level of preparedness
- Instrument flight
- Irregular/stressful work hours
- Jet lag
- Lack of sleep
- Mechanical vibration and noise
- Multiple flights or legs per day
- Night flight
- Physical discomfort
- Poor or inadequate diet and/or exercise
- Stress-induced anxiety
- Temperature extremes or changes

As pilot fatigue increases, the pilot's ability to fly decreases. In extreme conditions of fatigue, the pilot may become completely incapacitated and incapable of rational decision making. Brett was an example of how severe fatigue can affect a pilot. His condition was brought on by extended stressful work periods with inadequate rest. The effects of fatigue are amplified at night. A tired pilot will experience a greater desire to sleep at night than in the day. As metabolism slows down, the fatigued pilot will experience increased symptoms, possibly demonstrating lousy, irrational judgment and reacting slowly to potentially hazardous in-flight situations.

To prevent the onset of fatigue, a pilot needs only pay attention to personal health—get adequate rest, eat well, manage stress, and avoid harmful conditions/substances. If the pilot notices symptoms of fatigue in spite of taking adequate precautions, then something is wrong, and the pilot should not continue flying and should get assistance.

Vision

Your eyes could easily be compared to a camera. A camera gathers filtered light through the lens, controls its intensity with an aperture, and focuses the image onto a film of chemically treated plastic. The lens may be adjusted fore and aft, changing focal lengths to compensate for the varying distances of the objects photographed. The aperture works like a window of adjustable area, which can open wide, to admit maximum light, or constrict to barely more than a pin-prick in very bright conditions, necessary because chemicals on the film function only in controlled lighting conditions. (See Fig. 4-2.)

Light coming into your eyeball passes through the cornea (filter), is modulated by the iris (aperture), and is focused by the lens onto the retina (film). The basic principles are the same, but the eye is remarkably more capable than a camera in a couple of fascinating ways. First, the lens in your eye does not move forward and back, like the camera. Your eyes' lenses are like clear marbles that are soft and flexible. They are surrounded by a ring of fibrous muscle tissue, a little like the rings around planet Saturn. When the muscles pull on the lens, it is stretched thin, when the muscles relax, the lens thickens, thus changing the focal characteristics of the image on the retina. When you get old, that soft, pliable lens may become a bit stiff, making it difficult for you to see objects close-up. The muscles that adjust the focus of your lens may suffer from fatigue—anything that exercises your eyes for an extended period, such as reading a lot of fine print, will cause you to feel the effects of this strain, making your eyes feel tired. With this in mind, your eyes will relax, whenever they can, such as when you are sleeping. They will rest while you are awake, as well—particularly if there is nothing to look at, such as a night-blackened sky. Your resting eye will tend to focus at a range of about 4 to 6 ft away, and you won't necessarily be aware of it—a phenomenon called *empty-field myopia*. If you drone along in the plane at night, with no obvious details visible outside the plane, your eyes might relax, causing you to see a range no further distant than the windshield for hours at a time. You can easily solve this problem by simply looking at a distant object, such as the wingtip or a clearly visible light on the ground, and keeping the focus while you scan the view outside.

The second amazing feature of your eye is its reusable, motion picture film, the retina. This is where the eye is wonderfully complicated. The retina is surfaced with millions of tiny nerve endings, which are sensitive to light. Most can distinguish color, while other, somewhat more sensitive nerves see only the raw light intensity—basically black

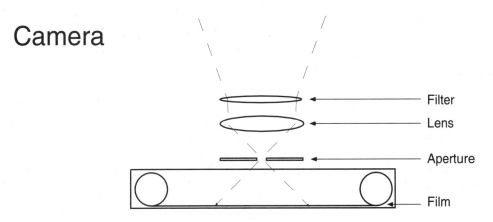

Camera

Filter

Lens

Aperture

Film

Fig. 4-2. *A camera is basically a mechanical eye.*

and white. Each nerve ending in the retina is individually wired to your brain and, when hit with light, can fire off a small electrical impulse into the heart of your brain's image processing center. These small electrical impulses are assembled in your brain to form an image, much the way a computer image is created from numerous pixels or a television image is formed from numerous small dots of colored light. (See Fig. 4-3.)

A closer look finds that the small nerve endings in your retinas contain a light-sensitive chemical which breaks down when exposed to light. The chemical reaction produces the electrical impulse in the nerve. Now, once the reaction occurs, the nerve is not sensitive to light any more—it doesn't "see." As with a gun, once the trigger is pulled, it won't fire again until it's reloaded. You can experience exactly this sort of phenomenon by having your picture taken in a dark room with a flash-bulb equipped camera. The flash goes off, and you see nothing but white for a few seconds—the light-reactive chemicals in

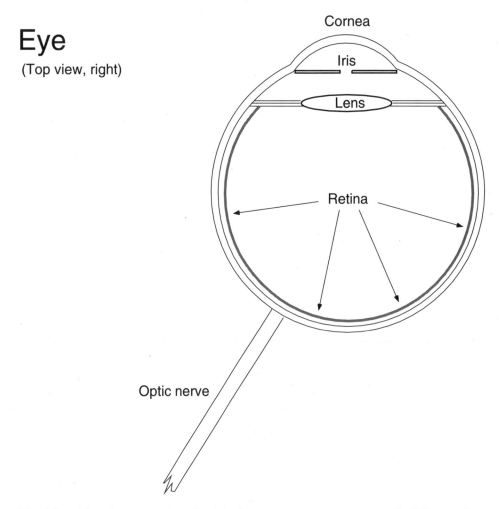

Fig. 4-3. *Although it appears simple—like the camera—your eyes are wonderfully complex.*

your eyes have done their job, and for a moment you are blind. But wait, in a few moments, your vision returns! What is going on? There is a massive reloading project occurring within the nerve endings in your eyes. The light-reactive chemicals are being re-built with some enzyme action and exposure to large quantities of oxygen. The blood-borne oxygen comes to your retinas through a large artery to the rear of each eyeball. The blood flow is impressive. If that artery were to break, you could bleed to death in a few minutes. As you sit, reading this book, your eyes are consuming up to 70 percent of the oxygen you breathe, tending to the chemical needs of the nerve endings in your retinas.

Here is where the iris comes into play. Its job is to regulate the light that enters your eye so that your retinas are able to keep up with the exposure. You might think of each nerve as a soldier in a firing line. The iris acts as a battlefield commander, desiring to cover the enemy with continuous fire and thus ordering some "soldiers" to shoot while others are busy reloading. As each gun is fired, it is reloaded while the others shoot. This is in contrast to everyone shooting at the same time, leaving the army momentarily defenseless while reloading. You have so many nerve endings in your retinas that the processes of exposing and recharging the light-sensitive chemicals in your eye happens at the same time, on a continuous basis, so that you may see a constant, moving image. The flashbulb simply surprised the iris, catching it wide open in dark conditions, and overloaded the nerves on your retinas, causing all the nerves to "fire" at once.

The nerve endings in your eye comprise at least two distinct varieties. Close inspection reveals that some resemble small cylinders, others are more conical. They have thus inherited the names, *rods* and *cones.* The cones contain chemicals that are reactive to light of different wavelengths, enabling them to respond to color. The rods contain a similar chemical which overall is much more light-sensitive but does not distinguish wavelength, or color. The rods, because of their sensitivity, have their greatest use in nighttime, low-light conditions. These different types of nerve endings are distributed about the surface of the retina, with cones favoring the areas in direct focus of the lens, rods more to the periphery. This is natural. Since your eye functions primarily in daylight conditions, it is optimized for seeing in rather bright sunlight. So, in daylight, the high density of cones at the central focusing area of your retinas, called the *fovea,* enables you to distinguish color and clarity unrivaled by the very best of camera equipment. Night is another story.

Since the cones are adapted for daylight conditions, they distinguish relatively little at night. The rods, at the periphery of your vision, are much more capable in darkness. The rods are also more sensitive to motion, enabling you to react to a flicker of movement just visible out of the "corner of your eye." At night, these rods provide the majority of the image you see. Unfortunately, the rods are most prevalent in the areas of your retina that correspond to peripheral vision, so your ability to see an object in the dark is actually better if you do not look directly at it. You can experience this phenomenon by looking at the stars in a night sky. Your peripheral vision will sense vast numbers of stars, including many faint details that seem to disappear when you turn to look at them directly. Your color vision does not become dormant at night, however. It still functions but requires relatively bright light to work. If a light source, such as a neon sign, is bright enough, your eye will distinguish it in detail and color, as in daylight. Under normal nighttime conditions, however, lighting is inadequate for the cones to function well, and you have difficulty distinguishing much beyond the muted grays and blues of darkness.

Your eyeball is colored black on the inside, like the black insides of a theater or camera body. The black pigment on the surface of your retina prevents light from glaring reflectively around inside your eyeball. Having no pigment would be a bit like watching a movie in a theater with the lights on. Albinos are born without such pigment in their eyes (which causes the red-colored eyes—you see the blood vessels) and suffer from poor visual acuity because of this. Believe it or not, vitamin A is contained in this pigmented surface, as well as within the cellular fluid of the rods and cones. With a little enzyme action, vitamin A can be converted into more of the light-sensitive chemical that allows your eye to work—it is already very similar in structure. This becomes a real factor in the long term, as your eyes have the capacity to adapt to a dark environment and increase their light sensitivity by producing more of the appropriate chemicals from vitamin A. You could compare this to putting a higher-speed film in your camera. Conversely, the chemicals in your eye can be converted back to vitamin A by simply allowing them to break down as they are exposed to light. Your eye can in this way adjust its sensitivity to the light conditions outside by varying the amount of light-sensitive chemicals in the rods and cones—literally changing the sensitivity of the "film."

If you insist on operating in a dark environment, your eyes will adjust their light sensitivity as best they can to enable you to see—a process called *dark adaptation.* It takes about 2 hours in near-total darkness for your eyes to adjust completely for nighttime. It's a common practice for military pilots to sit in the dark, wearing dark glasses, for a few hours before launching on a night mission—they want to see as well as possible. Any night pilot could benefit from this adaptive capability. Your eyes can adapt the other way, however, and they adjust to daylight conditions very quickly—in only a few minutes. It's a lot easier to break the sight chemicals down than to make them. A pilot who wishes to retain the best of night vision should keep cockpit lights very low and avoid looking at any bright lights (Fig. 4-4).

There is a condition called *night blindness,* which may occur when you have a deficiency of vitamin A. Like the name implies, you would be unable to see in dim-lighting conditions because your eyes do not have sufficient raw materials from which to create the necessary chemicals you need to see. Since vitamin A is usually stored in large quantities in your liver, it would take many months of a vitamin A–deficient diet before you would notice any symptoms in your eyes. Interestingly, night blindness can be reversed within an hour by an intravenous injection of vitamin A. For the pilot, however, the lesson is quite clear—eat your carrots.

All the nerve endings in you retina are bundled together into a single, rather thick optic nerve, which trails like an extension cord off the back of your eyeball and into your brain. Coupled to the optic nerve is the main artery and vein that supply blood to the eye. You could picture them as a garden hose and electrical cord attached to a beachball. There is enough slack in the bundle to allow for the articulation of your eyeball. The point on your retina where the optic nerve and artery attach is devoid of nerve endings, forming a small blind spot, called the *optic disk.* The specific location is a little off-center, to the inside of each eye. The blind spot in one eyeball is compensated by the image in the other, and vice versa. If you close one eye, the blind spot is visible, but your brain paints over the image, by inferring details from what is visible. You may "see" this blind spot with a simple exercise: Make a couple of small *X*s, about 2 inches apart on a piece of paper. (See Fig. 4-5.) Covering your left eye, look only at the mark on the left, adjusting the paper forward and back, until the other *X* disappears—its image is focused on your blind spot. Covering your right eye, look at the mark on the right. You'll witness the same phenomenon. An interesting variation of this is to draw a line connecting both the marks, extending it a little beyond each. (See Fig. 4-6.) Repeating the exercise, you will find that the mark disappears, but the line is unbroken. Your brain has simply inferred the existence of the line across your blind spot, and "drawn it in" for you. (See Fig. 4-7.)

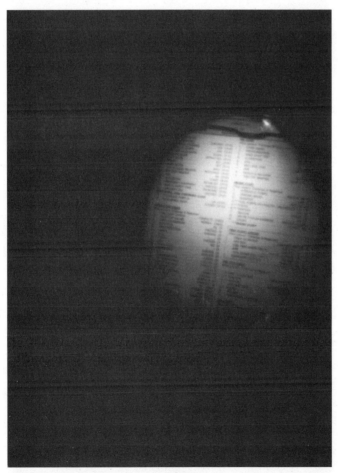

Fig. 4-4. *Use low light to read checklists and charts. Bright light will temporarily ruin your night vision.*

Fig. 4-4. *Use low light to read checklists and charts. Bright light will temporarily ruin your night vision.*

All of this wonderful anatomy is housed like a ball and socket, manipulated, aimed and controlled by three pairs of muscles attached to the outside of your eyeballs. These muscles function in a coordinated effort to aim your eyes at things you wish to look at. Typical for musculature in your body, the muscles that guide your vision tend to do so in leaps and jumps, called *saccades.* When you read this page, your eye is literally darting from one point to another on a given line of print, at a rate of two to three times per second, in an action known as *saccadic movement.* Your eyes fixate momentarily at a given point of interest, then leap to another, and so on. It happens so rapidly that less than 10 percent of the time is actually spent moving your eyes, the other 90 percent has them fixated on something. This has the potential of making the image you see appear as though your eye were resting atop an active jack-hammer. This image is stabilized at the visual processing center of your brain, so that you see a rock-steady image of the world around you. Your brain automatically stabilizes the image by reference to objects of known stability. The ground doesn't move, so your brain makes the image of the ground appear steady. If you remove all background images, perhaps looking at a single light against a completely dark background, there is little for your brain to reference to stabilize the image, and you may suddenly become aware of the jerky motions of your eyeballs. The light may appear to jiggle and dance, perhaps darting about like a firefly or even another airplane. This is called *autokinesis,* meaning that the im-

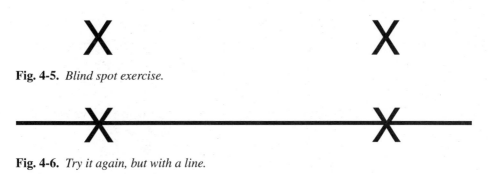

Fig. 4-5. *Blind spot exercise.*

Fig. 4-6. *Try it again, but with a line.*

Fig. 4-7. *Your brain is constantly "touching up" the images you see.*

age appears to have a motion of its own. If the horizon were visible behind it, it would just as quickly become stationary again, as your brain picks up on the reference. The system is utterly fantastic.

Depth perception is provided by the two, slightly different images that you see from each eye, as they come into the brain. It is very effective at distances ranging from close-up to about 30 ft or so. Beyond that, your brain compares perspective views and relative motions to perceive distance in the images you see. Subtle details, like the way mountains in the distance appear to move in relation to the stars behind them, would indicate to your brain that the mountains are closer than the stars. If many of these details are removed, as on a dark night, your brain may lose its ability to sense range and distance, making city lights, other airplanes, and stars look like a single, flat image, devoid of perspective. The same is true in landing on a dark runway; you might be surprised by how close—or how far away—the runway really is (Fig. 4-8).

Hearing, Balance, and Motion Sickness

The human ear is among the most sensitive of the animal kingdom. Dogs and bats can hear much higher frequencies, of course, and elephants and whales communicate at much lower frequencies, but as far as hearing acuity is concerned, there are few ears better than your own. The interesting aspect of your hearing is that your brain is forced, for its own sanity, to turn down the volume. At a given time, you are hearing a veritable barrage of background noise that, if you paid much attention to it, would cause difficulty in concentrating on whatever it is that you consider important. Take the pesky leak in the bathroom faucet, for example. You might not notice it at all in the daytime, when ambient noise and activity reduce its constant dripping to unimportance. If you lay awake at night, however, background noise quiets, and your still active mind begins to turn up the volume. You become aware of the chirping of crickets, the rustling of your blankets, and

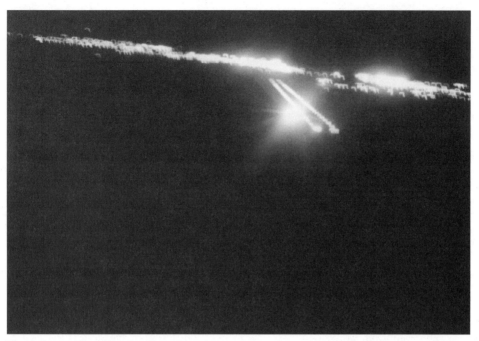

Fig. 4-8. *A "black hole" approach, to a runway like this one, offers little depth information.*

that incessant drip…drip…drip….All these sounds were present before, but you had the volume turned down.

Many years ago, as part of a school fieldtrip, my class was ushered into an echo-proof room at a university physics laboratory. Once inside, with the thick, insulated doors closed behind us, we were instructed to sit quietly and remain very still, for about 5 minutes. There was absolutely no background noise, as the room is completely isolated from all external sound. A little time passed and the student next to me retied his shoes—it was surprisingly loud. A little while later, I heard myself *blink*. We all could hear our own pulse, and even more interesting, the sound our own ears made, due to internal vibrations and blood flow. To some, it sounded like listening to a sea-shell, others heard a high-frequency tone. We were instructed to listen carefully to the street as we left the room. The sound of traffic was uncomfortable, and, from a good distance, we could hear the switches click as the traffic lights changed. It wasn't long before our hearing returned to normal.

When you fly, your hearing sensitivity is definitely turned down—airplanes are noisy. You may find that it becomes easier to sleep under such circumstances. Of your body's basic senses, hearing is a good indication of your level of alertness. Turn it down, and you may naturally relax. Couple this with the pressure changes due to altitude, and you may feel as though your ears are not working well at all. With the sensitivity of your ears turned down, you are still sensitive to low-frequency vibrations through your skeletal structure. Your bones can actually *hear*. You can demonstrate this to yourself with a little tap on your head—the sound it makes is carried to your auditory nerves through your skull. Sometimes these low-frequency vibrations that come from an airplane and literally rattle your bones can be extraordinarily fatiguing. Children appear to be especially susceptible to this. My 2-year-old son, for example, could hardly stay awake in a light plane after the engine started. This tendency toward relaxation and sleep could be aggravated at night, when your body wants to sleep, anyway.

Since your ears are rather delicate and may be damaged by continuous exposure to loud noise, you should use some form of hearing protection. Exercise care, when choosing a headset, to find a comfortable type; the discomfort caused by the pressure of an ill-fitting headset may contribute to fatigue. However, a good technique for staying alert is to periodically take the headset off. The sudden change in noise seems to help freshen the experience of sitting in a cockpit.

Aside from hearing, there are other sensations that your ears provide. It is your balance that is especially pertinent to night flight. Your sense of balance is determined by a mechanism deep inside your ears. The organ looks like a small, loose tangle of tubing—called the *membranous labyrinth.* It is composed primarily of three loops of tubing, arranged in half-circle arcs that join at their bases at right angles to align with the three spacial axes. Known as *semicircular canals,* these tubes contain fluid which runs about within them in response to movement. The rushing fluid triggers small, hairlike nerve endings whose stimulations produce a sensation of acceleration. Since the tubes are aligned with all three spacial axes, you can sense accelerations in any direction.

Your sense of equilibrium is based upon acceleratory forces, and, because of this, you can be fooled. You can easily become accustomed to steady motion, adopt a new frame of reference relative to that motion, and thus decide you're not moving at all. You are traveling at tremendous speed right now, for example, and you've never felt the sensation of it. Simply living on the earth means that you are traveling something near 1000 mph, because of its rotation. The earth itself is making about 66,600 mph around the sun, and the sun is really clipping along through the galaxy, and so on. You will forever be insensitive to these motions unless they were to change—then you would sense the acceleration or deceleration of the change. Someone once explained this rather succinctly by claiming that a long fall never hurt anybody—it's the sudden stop at the bottom. So you're sensitive to accelerations.

The limitations of your sense of balance are compensated by what you see. It is even possible for a person to maintain balance solely by visual information, but not particularly well:

> Many persons with complete destruction of the vestibular apparatus have almost normal equilibrium as long as their eyes are open and as long as they perform all motions slowly. But when they move rapidly or close their eyes, equilibrium is immediately lost. (C. Guyton, *Human Physiology and Mechanisms of Disease,* W. B. Saunders, Philadelphia, 1987.)

However, with normal equilibrium sense, it is simple for a person to maintain balance—and even move about—with their eyes closed. For pilots, a problem arises with the sense of balance. With eyes closed, it is impossible to fly an airplane for very long and maintain an accurate sense of upright. The airplane can create too many accelerations that mimic gravity but are not necessarily oriented toward the earth. A blind pilot, going by internal sense alone, can be easily fooled by normal aerial sensations into flying the airplane in almost any attitude while thinking the plane is straight and level. For this reason, the pilot's eyes referencing the horizon become the primary means of balance and orientation as far as the airplane is concerned. Since night causes some degradation in the pilot's ability to see outside, night may also affect the pilot's ability to keep the airplane upright. In other words, without the aid of attitude instruments, the pilot flying at night is a strong candidate for disorientation and vertigo. It is for this reason that night equipment should include at least some basic attitude instrumentation.

When the internal sensations of balance disagree strongly with what is visually indicated outside, the resulting vertigo can be nauseating. This may be the heart of motion sickness. The night VFR pilot might encounter some difficulty in seeing the horizon outside and thus subconsciously rely on, or pay more attention to, internal indications of balance. When the horizon is perceived, and the pilot feels a discrepancy with her or his internal sense of balance, a moment of disorientation and confusion may cause an upset stomach or other discomfort. Thus, the possibilities for motion sickness and disorientation are greater in the dark.

The FAA publishes an advisory circular on spacial disorientation (60-4A). Much of it is reprinted here, since it is short, direct, and to the point.

a. The attitude of an aircraft is generally determined by reference to the natural horizon or other visual references with the surface. If neither horizon nor surface references exist, the attitude of an aircraft must be determined by artificial means from the flight instruments. Sight, supported by other senses, allows the pilot to maintain orientation. However, during periods of low visibility, the supporting senses sometimes conflict with what is seen. When this happens, a pilot is particularly vulnerable to disorientation. The degree of disorientation may vary considerably with individual pilots.

b. During a recent 5-year period, there were almost 500 spatial disorientation accidents in the United States. Tragically, such accidents resulted in fatalities over 90 percent of the time.

c. Tests conducted with qualified instrument pilots indicate that it can take as much as 35 seconds to establish full control by instruments after the loss of visual reference with the surface. When another large group of pilots were asked to identify what types of spatial disorientation incidents they had personally experienced, the five most common illusions reported were: 60 percent had a sensation that one wing was low although wings were level; 45 percent had, on leveling after banking, tended to bank in opposite direction; 39 percent had felt as if straight and level when in a turn; 34 percent had become confused in attempting to mix "contact" and instrument cues; and 29 percent had, on recovery from steep climbing turn, felt to be turning in opposite direction.

d. Surface references and the natural horizon may at times become obscured, although visibility may be above visual flight rule minimums. Lack of natural horizon or surface reference is common on overwater flights, at night, and especially at night in extremely sparsely populated areas, or in low visibility conditions. A sloping cloud formation, an obscured horizon, a dark scene spread with ground lights and stars, and certain geometric patterns of ground lights can provide inaccurate visual information for aligning the aircraft correctly with the actual horizon. The disoriented pilot may place the aircraft in a dangerous attitude. Other factors which contribute to disorientation are reflections from outside lights, sunlight shining through clouds, and reflected light from the anticollision rotating beacon.

Respiration and Oxygen

Your body consumes oxygen constantly. Oxygen passes from the air in your lungs across the membranes in your lungs to your blood and is carried by the blood to your organs and tissues. The oxygen is literally combusted with carbon-based "fuels" derived from the food you eat. The burned carbon, CO_2 is absorbed back into your blood, passed back across the membranes in your lungs, and exhaled into the air. We've already discussed the critical requirement of oxygen in your eyes, but it is equally critical in other parts of your body. If a pilot were to deprive oxygen to the body's systems by pulling excessive "g's," for example, sending blood by the factored force of gravity to his lower extremities, his vision would fade to white—very similar to the picture flash, but in slow motion. Without oxygen to rebuild the light-responsive chemicals in his eyes, vision fades away. The pilot would be blind but conscious for a few moments; once the pilot's brain was sufficiently starved of blood, he or she would be unconscious. If the acceleration were to continue unchecked, the pilot would die from it in a few minutes. This, of course, rarely happens because the airplane, with an unconscious pilot, usually crashes before the "g" loading gets her or him.

Another way to deprive a pilot of oxygen is to fly at high altitude. With altitude increases, the pressure of oxygen in the lungs may be insufficient to cause it to pass across the membrane into the blood. Sudden depressurizations at very high altitudes can cause the oxygen pressure in the body to be greater than that of the surrounding air, causing the pilot to actually exhale oxygen—literally deflate and bring unconsciousness and death in moments.

Under more normal circumstances, the pilot is trained to observe symptoms of hypoxia, as they often occur slowly, gradually depleting the pilot's capabilities over an extended period at moderate altitudes. One of the effects of this hypoxic state is a reduction in visual acuity in the dark. The eyes need oxygen to function. Lack of oxygen will reduce the eye's sensitivity to low light. For this reason, the FAA recommends that a pilot consider using supplemental oxygen at altitudes above 5000 ft, at night. This is completely relative as far as the pilot is concerned. I have lived at an elevation of 4600 ft, or so, for the last 30 years. My body has acclimatized to this altitude. To use oxygen at 5000 ft seems silly for night operations in the traffic pattern . Nevertheless, that is my circumstance, and relative to my condition. Pilots from lower elevations might experience night vision degradation at lower altitudes, and the use of supplemental oxygen would certainly enhance night vision to some degree as altitudes increase.

Psychology

What's going on in a pilot's mind as the plane flies through the night? There are surely as many answers to that question as there are pilots; however, in general terms, there may be a few consistencies. One is a quiet state of denial. Many pilots feel that the night emergency won't happen to them—the engine will keep running, the weather won't close in, and so on. This is a normal response to potentially dangerous conditions which allows pilots to feel more comfortable: Pilots wouldn't fly if they knew their engines were going to fail. Not knowing, and hoping everything will be all right, the pilot goes flying. Denial is a form of optimism, which, when coupled with enough pessimism to inspire the pilot to double-check everything, is okay. The opposite of denial is paranoia, where the pilot feels that everything bad will happen, causing feelings that border on the irrational sense of panic. Neither mode of thinking, in its extremes, is safe. A pilot in severe denial will insist that no emergency will happen, even after one has already occurred. A pilot subject to paranoia might make an emergency response to a normal condition, creating a real emergency in the process. We should strive for the middle ground.

We have a need to be entertained. If you lose interest in what you are doing, doing it well becomes difficult. For pilots, flying the airplane can be boring—especially when the view disappears outside. Long stretches of night or IFR flight can cause a pilot fatigue and much mental stress simply because it isn't fun anymore. If the pilot were strongly fatigued at the same time, sleep may become an irresistible option. Sometimes daydreams will take on an uncanny reality. Lindbergh heard voices behind him, as though a whole group of observers sat in the tail of his plane. Later, he saw visions of whole islands and land-masses pass by below in excruciating detail, when nothing was really there. He was very fatigued, and his mind became distracted with its own entertainment. So many pilots have flown for long stretches, quietly wishing for something different to happen. I can remember flying around the traffic pattern in daylight for "umpteen" thousand times, secretly wishing we'd have an emergency of some kind, just to break up the monotony. Wishing for an emergency? How foolish! And yet these irrational thoughts can arise when the pilot is bored. Night flight, especially for long distances over water, can be exceedingly boring. You might bring along some entertainment, the best of which is another person in the cockpit. If you are alone, find things to occupy your mind with the task of flying—try to enhance the airplane's performance, improve the operation of its systems, eat, listen to music, or strike up a conversation with a controller.

Although many pilots report that night flying is relaxing, most report higher levels of anxiety about the possibility of an engine failure. This anxiety stems from the fact that an in-flight emergency in the dark could require an especially difficult landing; indeed, the odds of surviving an engine failure in darkness are about half as good as in daylight, and a powerless descent into blackness is a frightful prospect for anyone. With this in mind, pilots tend to fuss over the engine operation a bit more than in daylight, watching and listening closely for any sign of impending failure. A natural result of all this attention given to the engine is that the pilot may notice unfamiliar sounds or vibrations. These anomalies have, in most cases, been there all along, but flying at night with a heightened sense of anxiety elevates the pilots level of awareness and the sounds are suddenly manifest. This is called *automatic rough* and is a perception of roughness caused by the pilots anxiety over an engine failure. Anxiety levels, conscious or not, can be expected to increase in darkness. They can affect how a pilot responds in an emergency, possibly lowering the panic threshold and causing the pilot to make irrational decisions.

However, the invisibility of much detail may provide a simplified view of the night environment. Occasionally, the only things obviously visible have direct bearing to flying the plane, such as the airport runway and other air traffic. The fact that extraneous details are not visible or distracting to a pilot's attentions gives rise to the impression that night flying is relaxing (Fig. 4-9). This may seem like a contradiction of the above paragraph, but when coupled with denial, the night pilot may easily be lulled into a strong feeling of satisfaction and wellbeing. The simplicity of the view at night is why most flight simulators favor night environments—it requires far less computer memory to produce the night image than a VFR daylight image. Perhaps for similar reasons, the night pilot feels that less effort is required to keep track of the flight environment in darkness. When an emergency actually occurs, however, the shock of it can be amplified, as the pilot goes from a relaxed state to one of rather severe alarm.

Fig. 4-9. *The view in darkness seems beautifully simple, but an emergency landing in invisible terrain might be a fearful proposition—this city is surrounded by mountains.*

Chapter Four

Judgment

The pilot's ability to make decisions can be affected by darkness. Consider this example. The pilot flying a single-engine airplane at cruise altitude is faced with an engine failure. His heart jumps in his throat. His reverie in the cross-country trip is suddenly shattered by the whispered stillness of a prop windmilling outside. The pilot mentally goes through the three steps of an engine failure: Fly the plane. Pick a place to land. Attempt a restart, if there is time. The pilot begins to slow to best glide speed. Looking outside, there is little to see. The pilot imagines all sorts of horrible features on the ground below, any of which could cause a fiery crash. Flying the plane straight ahead, the pilot focuses attention on the engine. It has to work. It *has* to. He works the throttle, the mixture, switches mags, tanks, checks carburetor heat. The engine doesn't respond. As the plane descends, the pilot still hasn't selected a place to land. His anxiety is increasing rapidly now—he's becoming irrational. Since the pilot cannot visualize the emergency landing, he becomes focused on getting the sick engine to run again. As his attentions narrow on the engine, flying the plane becomes less important, and its glide steepens; the pilot has not given much thought to the direction of flight. Busily working with the engine controls, now approaching panic, the pilot has practically left the plane to fly itself. And that's the way he dies, hunched over the engine controls, pulling, switching, and tugging, while the plane flies over numerous survivable landing sites and crashes, uncontrolled, into the ground. Sadly, this scenario has been duplicated many times.

The pilot cannot afford to become irrational. Almost all the physiological aspects of night flight can affect the pilot's rationality. Of these, perhaps the greatest is fatigue, but the others should not be severely discounted. Stress levels can accumulate rapidly during critical night maneuvers. At many rural airports, the simple act of flying the pattern can approach the complexity of an IFR circle-to-land procedure, requiring the pilot to divide attentions rapidly from instruments to outside, to configuration changes and the radio all at nearly the same time. In extreme situations, such as the simultaneous occurrence of an emergency or visual illusion, the pilot's capacity to perform might be surpassed. In Brett's example, his capacity was saturated by the simple act of trying to stay awake—it occupied his full attention and grew in severity until he almost gave up the fight.

Consider a juggler at the circus. She is comfortable juggling, say, three balls. While doing this, she can entertain the crowd with jokes and alter her manner of juggling to bounce the balls and adjust their order in her hands. Suddenly she adds another ball. With four, she moves faster, touching each ball for an instant as she sends it flying to the other hand. At five, her face is a picture of concentration, her hands a blur. She has stopped talking and begun to sweat. This is like a pilot during a night landing approach into a difficult airport. The juggler is still there, but an enterprising clown adds a chainsaw to the fray. Now the juggler is really busy—more than that, she is saturated. No matter how skillful a juggler she is, there is a limit to the number of items she can juggle, especially the pressing ones, like the chainsaw. It is the same with pilots. No amount of skill can put a pilot above the possibility of becoming saturated with tasks. With the addition of a chainsaw, the juggler drops one of the balls. Her capacity is five—period—and that will change from day to day, depending upon her condition. The chainsaw required that she lighten her workload, so she dropped something relatively menial. Pilots at night need to prioritize their tasks, as well. For the hapless pilot who crashed while trying to start the engine, he focused on the menial task and was killed by an important one. A pilot demonstrates good judgment when personal limits are well understood and the situation managed to prevent exceeding them, for a pilot who "drops the balls" dies.

The effects of night can easily reduce your capacity to fly the plane, effectively lowering your saturation levels. For example, daylight VFR flight provides an easy reference horizon with which to reference the plane's attitude. In that condition, you might easily cope with a number of factors including some nasty emergencies. Nighttime, on the other hand, might force you to handle the problem while referencing instruments, limit your options of landing sites, and simultaneously raise your anxiety levels. You could reach your saturation level much sooner.

TECHNIQUE

Your physiological needs are always a factor when you fly a plane. They cannot be ignored. The point of this section is to highlight a few preparations you should make to help you cope with your body's needs in darkness.

Eyes

If you plan to fly in dark conditions, you can maximize your vision by sitting in darkness for at least 2 hours before you go fly. Artificial darkness works fine. You may shut drapes, close the door, and wear sunglasses with the lights off. Listen to music or sleep. Once your eyes have fully adapted to darkness, it becomes important to avoid light conditions. If you see bright lights, you will destroy the past few hours' effort in a couple of minutes. On the flight line, avoid looking directly into landing lights. Since this is sometimes unavoidable, or perhaps closing your eyes would be foolish, close one eye to protect at least half your night vision. Use red lights in the cockpit. Studies indicate that red light does not alter your night vision yet still provides enough illumination for you to see charts and instruments.

When looking for obscure details outside, try not to look directly at the object you wish to see, but look to one side of it. This will put the image on a part of your retina that is optimized for night vision, enabling you to perceive the object with greater clarity.

Sleep

It may seem obvious to suggest that you get adequate sleep in preparation for a night flight, but the practicalities here can easily prevent it. Your body is probably not accustomed to sleeping in the day, preventing you from sleeping at all. The demands of work and other daylight time constraints might also interfere with your opportunities to take a nap. You must consider all these factors and manipulate your schedule to allow for adequate rest before you fly off into darkness. Pilots who make regular night flights find that it takes many days, even weeks to train their bodies to sleep adequately in the daytime. If you consider a long night trip, you had best take advantage of a sleep opportunity beforehand. Ignoring your need for rest might severely hamper your ability to fly—it could even kill you.

When you fly, ask yourself honestly, how do you feel? Be prepared to skip the flight if you don't feel up to it. Brett knew that he shouldn't be flying, but to help the company, he flew anyway. It was a good time to simply say no. Look your mental state over carefully, discounting what might be normal excitement to be going flying. Ask yourself these questions:

- Have you had enough sleep?
- Have you slept recently?
- Are you mentally tired?
- Stressed about anything?
- Are you angry?
- Are you in a hurry?
- Feel overloaded?
- Hungry?
- Healthy?
- Does the flight make you nervous?
- Are you scared?

You might make up several questions of you own. The point here is that you honestly assess your own condition and make a decision about your ability to fly. If you have doubts, don't go. If you're not sure about some aspect of your condition but can't place your finger on exactly what, you might seek a friend to go along, one who is competent in the airplane.

You might use coffee or other caffeinated beverages to help you stay awake at night. That's fine, as far as staying awake is concerned, but you will need to make some preparations for your bladder. The pit-stop options are numerous: You could plan a flight with shorter legs, allowing you the opportunity to visit the bathroom periodically. You could bring along a urinal, although the opinions vary widely on the effectiveness of these items, especially among women. *Aviation Consumer* published an article describing how their staff sampled the use of various brands of urinals and "in-flight range extenders" in effort to find out which varieties worked best. Their findings were interesting, to say the least, ranging from delight to rabid hatred. Surprisingly, the item that seemed to interfere the least with the task of flying the plane was a simple diaper. Most airline flight crews have the benefit of an on-board bathroom. I work as a pilot for a regional airline. During one night flight into Rapid City, South Dakota, the captain, who drank several cups of coffee, got up to use the restroom and left me alone on the flight deck. He was gone for a long time. I had already begun the descent into the airport area, and wondered what had become of the him. About then, I got a call on the interphone. It was the captain. "Open the %*@! door! Didn't you hear me knocking?!" He'd locked himself out of the cockpit. He struggled with the door a minute, in full view of the passengers, then had to ask the flight attendant how to work her phone. I let him in, but I think the passengers heard me laughing. You'll need to plan for the bladder contingency.

Staying Awake

In flight, the techniques pilots use to stay awake are numerous. Coffee is one, but eating carrots or chewing other noisy, crunchy food works well, too. I mentioned sucking on ice-cubes in a previous chapter. Breathing oxygen can help you stay awake. Listen to the radio, talk to yourself, sing, chew gum, slap yourself, clap your hands, adjust your seating position, maneuver the plane, fly lower, read a chart, write down the indications of each of the plane's instruments, make a performance log, find out how far it is from your position to the nearest big city or how far you are from Nome, Alaska, and, if you must, turn up the interior lighting.

Exercising in the cockpit helps keep you alert, as well, by elevating your heart rate and giving you something to do. On the down side, it can also help you relax—so when you finish exercising, find something else to occupy your attention or simply plan to exercise for the rest of the flight.

There are several exercises that you can do while sitting and flying an airplane. Here are some examples:

- Start with your neck. Tip your head back as far as it can go, stretch your neck muscles, then tip your head forward—this is not to be confused with nodding off—to stretch the back of your neck. Do the same thing from side to side. Move slowly and hold the extreme positions for a few moments. You should feel some stretching in your neck muscles. Don't forget to keep flying the plane, and move slowly enough to prevent getting vertigo.

- Move down to your shoulders. Rotate your shoulders around while keeping your hands where they are. Move slowly at the limits of your shoulder's travel. You should feel some stretching here, as well.

- With each arm in turn, reach as far behind you as you can, then over your head, then to the floor. With the opposite hand, pull your elbow across your chest as far as it will go.

- Clasp your hands together and stretch them both over your head and behind you as far as you can reach. Clasp your hands together in front of your chest and pull as if you were pulling them apart, hold the pressure for a moment, then relax.

- Press your hands together as though praying and push hard, holding the strain for a few moments before relaxing again.

- As you sit, twist your torso from side to side, holding your elbows together in front of your chest while moving them from side to side, with your hips stationary.

- Keeping your back against the seat, reach one elbow at a time toward the bottom seat cushion, stretching your torso from side to side.

- Arch your back, pressing your shoulder blades against the backrest and pushing your stomach out as far as it can go. Reverse the motion by shrugging your shoulders forward, bowing your back against the seat and sucking your stomach inward.

- Take a deep breath and hold your breath as long as you can, then exhale. Repeat.

- Press your hands on your seat and lift both feet off the floor, such that your thighs are not supported by the seat cushion. Hold that position for a while, then relax. Do it again.

- Loosen your safety belt, press your hands on the armrests or seat cushion, and lift your whole body off the seat. Hold that position for a time then relax.

- Lift each leg individually, bend and straighten the knees, roll your ankles around, and make a circling motion with your feet.

- Stomp your feet on the floor several times.

- You might also soak a towel in ice water and put it on your face, over your head, up your shirt or in your lap. Pour cold water on your hair. Open the cockpit vents. Keep cool. It is more difficult to sleep when you are just cool enough to be uncomfortable while resting.

With all the commotion you might be making in the cockpit, don't forget to fly the plane. If you have a companion in the cockpit who is competent in flying the plane, caution him or her to stay awake, then make yourself comfortable and take a nap. If you are alone and are having difficulty staying conscious or even interested in flying, you'd best pick a nice airport and land the plane.

Vertigo, Disorientation, and Motion Sickness

One of the finest students I ever had was a victim of motion sickness. Try as we could, we couldn't shake it. Early on, during his training, it became evident that only short flights would do. We would fly until his stomach became upset, then discontinue the lesson and try again. He learned to fly with a smoothness the likes of which I haven't seen since. We needed two or three lessons to get through stalls. On his solo cross-country flight, there was enough light turbulence to make the poor guy fly with one hand and hold a bag over his mouth with the other. Over the course of the training, his condition improved a bit but did not go away. Thankfully, we were careful enough that he never actually vomited during any lesson, until the very end—a night flight.

He made nine beautiful landings at an outlying airport. On the way back, we played with the airplane, making some steep climbs and turns. He had fun until his stomach began to throw in the towel. We immediately settled down and flew toward home. The approach controller advised us of oncoming traffic, 500 ft lower, against the city lights. We'd been able to locate all the earlier traffic but had difficulty finding this one, disguised among the lights below. We discussed that for a while, until I noticed the airplane passing below, out my side and down. It was still quite difficult to see, even 500 ft away with its lights blinking. I banked the plane to the right, to point it out to my student. He saw the plane, but when we rolled back to level, he broke sweat....

"Quick, get the bag!"

"Where is it?" I said.

"In the back....Hurry!"

I couldn't reach his flight bag, it was positioned too far aft in the baggage area of the C-152. I quickly unbuckled my safety belt and began climbing over the seat...

"Quick, take the controls!"

I flopped back into my seat and took hold of the yoke. There was a blast of air as he opened the window and leaned outside. A vague smell of vomit permeated the cabin. I felt bad about banking the plane, and began to apologize.

"Hey, no problem, it wasn't bad—but we're gonna' have to clean mashed potatoes off the airplane."

This student went on to pursue the instrument rating but was unable to continue because he would get sick on almost every flight. He needed to see the horizon in order to keep his stomach settled. This is common. Lack of outside reference seems to contribute to motion sickness. If you plan on flying at night, it would be wise to position a sick bag somewhere handy in the cockpit.

The causes of vertigo were discussed in the previous section. The cures are simple. If you become disoriented, stare at the horizon or attitude indicator and fly by that reference until the vertigo goes away. If you don't focus on a realistic reference like that, the effects of vertigo tend to be prolonged. Vertigo generally strikes while the plane is turning and the pilot's head moves simultaneously. You can avoid getting vertigo in the cockpit by making your movements slow, particularly your head—where your inner ear is located. While reaching for items or controls on the floor, such as the fuel selector, try to keep your head upright and an eye on the attitude indicator.

There are other ways of getting disoriented. Getting lost, for example. The ease with which you can get lost at night is highly dependent on your geographic location. Flying over the ocean presents few lighted references, and the few ships that are there tend to move. Certain areas over land can also be difficult, but they are few, and, conversely, in VFR conditions it is often possible to see your destination from tremendous distances at night. So, generally speaking, you will probably get lost more often in daylight than at night.

The following is from AC 60-4A again:

a. You, the pilot, should understand the elements contributing to spatial disorientation so as to prevent loss of aircraft control if these conditions are inadvertently encountered.

b. The following are certain basic steps which should assist materially in preventing spatial disorientation.

(1) Before you fly with less than 3 miles visibility, obtain training and maintain proficiency in aircraft control by reference to instruments.

(2) When flying at night or in reduced visibility, use your flight instruments, in conjunction with visual references.

(3) Maintain night currency if you intend to fly at night. Include cross-country and local operations at different airports.

(4) Study and become familiar with unique geographical conditions in areas in which you intend to operate.

(5) Check weather forecasts before departure, en route, and at destination. Be alert for weather deterioration.

(6) Do not attempt [VFR] flight when there is a possibility of getting trapped in deteriorating weather.

(7) Rely on instrument indications unless the natural horizon or surface reference is clearly visible. You and only you have full knowledge of your limitations. Know these limitations and be guided by them.

SKILLS TO PRACTICE

Everyone manifests different symptoms of fatigue. This is an exercise to help you become aware of the symptoms that are yours. For this exercise to work, you need to get tired. Pick a long day at work, or if you want, plan to stay awake for a night and day. Early on, when you are refreshed and alert, roll dice 25 separate times and total the figure in your head. Have someone time you and verify your math. At the end of the long day, repeat the task and compare your performance. You may find that you are unwilling to perform the task the second time, that it took much longer, or that the whole process was irritating. Take note of the way you feel. In the future, if you feel that way in the cockpit, your performance is probably suffering and you should land.

Try this experiment with your vision. Set your alarm to go off early in the morning, while it is still very dark. Without turning on a light, wake up enough to read the headlines of a newspaper, or even its finer print, if you can. While you were sleeping, your eye became adapted to darkness—large quantities of photoreactive chemicals exist in your eyes. Then turn on the light and look at it until you don't have to squint anymore—you just caused your eyes to adapt themselves to light—the chemicals have been changed back to vitamin A. Turn the light off and try to read the paper again. I'd be surprised if you could see it at all.

FURTHER READING

FAA Advisory Circular AC 60-4A, U.S. Department of Transportation, February 9, 1983.

5
Take-off and climb

During the late 1940s, Wendover didn't officially exist. It ranked as one of the most carefully guarded secret air bases in the country (Fig. 5-1). It was from this base on the western edge of no-man's-land that a heavily loaded B-29 could practice dropping an atomic bomb. Since then, the area has eroded into a desert wasteland, leaving behind a vast, somewhat broken, concrete tarmac and several huge dilapidated hangars. There is some graffiti painted on the roof of one of the hangars, left there by some less-than-enthusiastic corpsman—"Welcome to Bendover, UT." The airplanes are mostly gone, except for the occasional visit of a regional airliner and a few transient aircraft. The tower sits unused and wind-swept, its once brightly painted red-checkered scaffold now fading to rust. Tumbleweeds blow by, rolling forever unobstructed to disappear into a shimmering gray horizon.

To the east there is nothing but salt and mud for 60 mi—the Bonneville salt flats, scene of numerous land speed records and innumerable bogged down cars. Interstate 80 cuts a straight swath across the wasteland, paralleled by a railroad track. North and south of the highway are huge areas restricted to military use. Sometimes, at night, you can see the flashes of explosions, and occasionally hear a sonic boom. A California artist felt inspired by the emptiness of the scene and planted a cement tree which looks like a six-story telephone pole suspending some ugly Christmas ornaments. It sits by the highway, in the middle of nothing, sun-baked and solitary. No one has understood

Fig. 5-1. *During World War II, this place was a carefully guarded military secret—and home base for the Enola Gay.*

the concept, and passers-by use it as a rest- stop, watering the side of its base that faces away from the road.

   To the west of the airport, and barely a stone's throw away, is the Utah-Nevada state line. The demarcation stands out by virtue of its sudden opulence—casinos, golf courses, tourist traps, and, surprisingly, mountains spring from the wasteland. It's not a big town, but the casinos work hard to be attractive to passers-by, spanning the highway at the State Line (a casino), promising cheap food and the best of luck. A four-story electric cowboy waves at passing motorists. The casinos and their lights make a brightly lit blip in the middle of a vast emptiness. The airport, far from its original clandestine glory days, is now simply the nearest runway to the proverbial $100 hamburger. It's a place for pilots to go, convenient to huge runways, and the casinos offer a shuttle service to the airport....

   Grant rented a Piper Seminole and departed Salt Lake City to fly into a beautiful desert sunset, happy to be logging some multiengine time. He'd invited me along for the ride. We divided the cockpit tasks—I managed the radios and trimmed the engines while Grant flew (it was his nickel). We finished eating well after dark and elected to walk back to the airport. With the lights of Nevada behind us, walking into the night of the eastern desert horizon, the blackness felt overpowering. We could see by the light of the airport beacon, and the occasional passing car, but little else. We could feel the night breeze and hear our footsteps but often couldn't see our feet. Without moonlight, the sky was liter-ally covered with stars. Looking up without visible ground perspective made me feel like

I was falling into space. The airport was completely silent in the darkness—the airplane dimly lit in regular intervals as though by the light of a prison tower. As we walked across the ramp to where the plane was moored, we felt the unseen vastness of the tarmac but could only see its rotating beacon, piercing the night like a lighthouse (Fig. 5-2). Sound carried for miles across the flat land. We could hear traffic on the highway, distant conversations, and the light breeze. I've always loved quiet airport ramps at night.

We made a quick check of the plane and set out to find the runway. There are no lights on the huge ramp, several chuck-holes, occasional coyotes, and a few rabbits. We found a faded yellow line, clicked the mike a few times on CTAF and observed runway lights a mile or so away—the yellow line would hopefully lead us there. Since the Seminole has rather limited propeller ground clearance, we taxied very slowly to avoid rubble. I know several people who were surprised at this airport when the ramp suddenly ended in the brightness of their landing lights and they roared off into the sage-brush, taxiing too fast, unable to stop in time. Another pilot got so frustrated trying to find a path to the runway in the dark that he just opened the throttle and departed from the ramp—it's certainly big enough, in places. We found runway 21 after about 15 minutes of looking, completed the run-up, and took off, full-throttle into the black night beyond the runway lights (Fig. 5-3).

When the plane flew, climbing past the end of the runway, there was nothing left to see. We were pointed away from any civilization, as if looking into outer-space. The windows were as black as if they had been painted, even though the visibility outside

Fig. 5-2. *Nowadays, the field is practically deserted.*

Fig. 5-3. *Departing into the dark void beyond the runway lights is like leaping off the edge of the earth.*

was unlimited—there just wasn't anything out there. Grant maintained attitude by reference to instruments and began a left turn for Salt Lake City. I busied myself with the throttle console and adjusting the finicky mixture on the right engine. I had everything set just about right when Grant surprised me with a sudden demand:

"Feather the left engine, we've lost it!"

"We have not! Don't touch the throttle. See? It's running perfectly."

"Then why am I holding all this right rudder?"

I looked up to see a beautiful string of lights—cars on I-80—just off the nose. The only thing visible as the plane turned, it was a ringer for the horizon. Grant naturally aligned the wings to the road as he rolled out of the turn, setting up what in reality was a 15-degree bank to the left—still turning. He was deeply convinced that he saw the horizon in those lights. Still watching his heading, he noticed a left turn on the gyro and dutifully compensated for it by pushing on the right rudder pedal. Even though the attitude indicator clearly showed the bank, Grant discounted its reading in favor of the view outside. He became alarmed when he had to push the pedal so hard, to prevent the turn (he'd have to, with 15 degrees of bank). Since he thought the wings were level, he concluded that an engine must have failed, and set forth to feather it.

With my eyes on the panel, I wasn't subject to the illusion of the false horizon as Grant was. I thought it was funny.

"Look at the attitude gyro, Grant—'ya weenie. You're in a bank to the left!"

"Oh. Oh, wow!"

He rolled the wings level, and I double-checked the engine instruments. When I looked up, he had the plane in a steeper bank to the left and had begun to lower the nose. As the plane flew toward the highway, the "horizon" appeared to move lower on the nose and roll to the left, giving the sensation that the plane was pitching up and rolling right. Grant could not break the illusion even though he was now conscious of it, and could not help but roll the airplane further left and lower the nose. We began descending—it was the descent that caught my attention again. When I looked at Grant, he was actually *leaning in his seat.* If it weren't for the cabin wall, I honestly think he would have fallen out of his chair.

"Grant, look at the attitude. You're banking again!"

We had not been climbing for very long, and didn't have much altitude to work with. I became somewhat alarmed. Grant had vertigo—bad. We had to break his faulty perceptions quickly (I was amazed at how tenacious they were), or he would be unable to fly the plane.

"Want to wear the hood?"

He declined, but needed some help for a short time thereafter, until we passed over the road and its illusory aspects faded away, no longer occupying his attention. Grant is instrument-rated, and the plane was well equipped, yet he had tremendous difficulty ignoring the illusion of the "horizon," to the point that he believed the instrument indications to be wrong. Without help, I believe he may have crashed.

BACKGROUND INFORMATION

For the most part, night departures are relatively easy. With a lighted runway, rotation and climb are not particularly difficult. If the horizon is visible, the departure and climb continue normally, as in daylight. The difficulty arises when these visual cues fade away or are misleading. Even under startlingly clear skies, the visual references in sparsely populated areas (no lights) can place an unsuspecting pilot in serious IFR conditions. In such cases, vertigo and death may occur within moments of leaving the runway. The effects of visual illusions and vertigo can be almost overpowering. For Grant, he lived to become a truck driver. Others haven't been so lucky:

Accident occurred April 14, 1990 at WENDOVER, UT

Aircraft: PIPER PA 24-250, registration: N6401P

Injuries: 2 Fatal, 2 Serious

> The pilot stated he activated the runway edge lights prior to taxiing for takeoff, and taxied for "quite a while" because the taxiway is not lighted. He believes the lights went out shortly after the airplane became airborne. The runway lights remain illuminated for approximately 15 minutes after activation. The surviving passenger remembered the airplane "tumbling" after takeoff. The airplane struck the ground in a left wing down, nose low attitude. The airport is located southeast of the city, and pilots familiar with the airport described the area along

the extended centerline of runway 21 as a "black hole." The pilot had logged 4 hours of night time in the last 90 days.

Probable cause:

The pilot's loss of control of the airplane due to spatial disorientation. Contributing factor(s) was: The lack of surface reference while taking off over a sparsely populated area under dark night light conditions.

Accident occurred March 13, 1994 at WENDOVER, UT
Aircraft: BELLANCA 14-19-3A, Registration: N8885R
Injuries: 4 Fatal

Area off the south end of the departure runway provides no ground reference at night. Shortly after take-off in dark night condition, the airplane collided with the flat dry lake bed, one mile off the end of the runway. The airplane collided with the terrain in a left wing low attitude. During the post crash investigation, there was no evidence found to indicate a mechanical failure or malfunction.

Probable Cause:

Inadequate in-flight planning/decision. Factors to the accident were: Clearance was not maintained and dark night conditions.

Wendover is famous for this type of accident. At least one FBO on the Salt Lake City airport won't rent a plane for a night flight that way unless the pilot is instrument-rated and the plane properly equipped—regardless of the weather. Many other airports offer similar conditions. Most of these are surrounded by large, uninhabited areas or bodies of water. Night departures from these areas require the pilot to give special attention to instruments and be well able to use them, even when the weather is ceiling and visibility unlimited (CAVU). Spacial disorientation on departure is only part of the challenge, however, as there are many ways to "go bump in the night."

Is the Runway Clear?

It's easy enough to taxi slowly when you can't see forward very far but much harder to take off in slow motion. Very often, there is no guarantee that the runway ahead is clear, especially at uncontrolled fields (Fig. 5-4). There could be almost anything out there beyond the beam of your landing-lights, including potholes, gravel, cattle, deer, dogs, wrecked planes, fallen trees, water, people—you name it. I know one guy who used to make a game out of riding his bicycle up and down the active runway at Salt Lake City International airport when he was drunk—and that place has a tower! You could hope that the noise and lights of your airplane would scare away any living obstacles, but many animals freeze when they are scared, and a cow could do wonderful damage to your airplane.

Can You See Lights?

You may be surprised to learn that runway lights are not legally required for your night operations (Fig. 5-5). Navigation lights for your airplane are specified in the regs, but landing lights are not, unless you operate for hire. So, if no one is paying for the flight,

Fig. 5-4. *How can you be sure the runway is clear?*

you could take off into the darkness with little more than navigation lights from an uncontrolled, unlit field. If you crash, however, the FAA may find legal justification by stretching the "careless or reckless operation" clause in the regulations to condemn an unlit night departure—and then the Drug Enforcement Agency might become suspicious, thinking that you have something to hide.

Attitude Control after Rotation

The horizon is the primary reference a pilot uses for attitude control. During the moments when the airplane accelerates down the runway, lights and peripheral indications often provide adequate attitude reference. But shortly after rotation, when the airplane has begun to climb, normal ground references become insignificant, and the pilot needs to quickly transition to other horizontal references. If the horizon is visible, great. If not, the pilot must take primary reference from the aircraft attitude indicator (Fig. 5-6). At Wendover, during Grant's dizzying departure, lack of an attitude indicator would have likely been fatal.

Mountain airports are famous for the need to avoid obstacles as part of the visual departure procedure. The proximity of tall obstacles precludes many of these airports from IFR operations. At night, the obstacles are usually unlit—mostly mountains and trees—and might be impossible to avoid. If you intend to use such an airport, daylight could be mandatory.

Fig. 5-5. *Runway lights may not be legally required.*

Transition to Instruments

The transition from outside to inside references can take a few moments for even the most seasoned instrument pilots. There is far more to it than just refocusing your eyes. As the airplane climbs away from the runway, the pilot must begin to reference the attitude indicator and commence a normal instrument scan. Often, the scan is a composite of external and internal queues. For example, the departure from Wendover requires internal attitude references but external navigation (for a VFR departure). The highway is still the most effective direction finder—it points directly at Salt Lake City, effectively marking the corridor between restricted areas, and the pilot is still responsible for "seeing and avoiding" other air traffic. It was this sort of composite scan that allowed Grant to be affected by the false horizon illusion. His mistake was to disregard the aircraft attitude indicator. If external queues are very sparse, the pilot should consider using an IFR departure procedure, hopefully anticipating this need and setting up navigation equipment in advance (Fig. 5-7).

TECHNIQUE

The goal of a successful departure is to accelerate to flight speed and climb to a safe altitude. In darkness, the pilot must do this with less visual information.

Fig. 5-6. *Even though visibility is better than 80 mi and the skies above are clear, a departure from this runway is a plunge into dark IFR conditions.*

Clearing the Runway

If you're not sure whether the runway is clear, taxi down its full length. This is a simple matter at uncontrolled airports where the pilot has the option of back-taxiing a runway to its end, then reversing direction for take-off (Fig. 5-8). The hope is that animals will not venture out onto the runway very soon after the airplane has passed by. A variation of this principle is to allow someone else to depart first, thereby ensuring that the runway is clear.

Avoiding Obstacles

Since obstacles near the runway may be invisible at night, the only clear avoidance procedure is to climb—fast (Fig. 5-9). Initially, the pilot should make use of a best-angle climb speed, gaining as much altitude as possible over known territory, like the runway itself. It may be wise to climb while circling over an area of known terrain before proceeding over high, unlit ground. For example, pilots flying east out of Salt Lake City at night often climb over the city lights until reaching an altitude well above the highest mountain on their planned route, before proceeding on course. This is similar to an IFR procedure, which would substitute a radio fix for the city lights and do the same thing without external reference. Without a city to fly over, many pilots will alter their courses to climb over roads or rivers to avoid invisible rock and forests.

Fig. 5-7. *On some night flights, this might be the only visible attitude reference.*

Use of a Climb Profile

Many aircraft types are uncomfortable at best-angle-of-climb speeds for very long. The typically low airspeeds and high power settings may combine to produce high engine temperatures. The high deck angle during the climb presents a problem in forward visibility, and for many pilots the speed itself is uncomfortably close to stall. With this in mind, a more practical departure might employ a variety of configurations, power settings, and speeds to enhance climb performance and obtain a good safety margin on departure. A horizontal, or profile, view of this climb path will often describe several steps where configuration changes and accelerations take place as part of the climb procedure. The entire procedure is known as a *climb profile.*

A typical climb profile in the Canadair Regional Jet as used by one local airline involves a configuration set to 20-degree flaps and rotation off the runway, retracting the landing gear, and an attitude such that the airplane climbs at single- engine best angle of climb plus 10 to 20 knots until 400 ft, a flap retraction to 8 degrees maintained until 1000 ft, flaps retracted and acceleration to best rate of climb until 1500 ft, whereupon the thrust is reduced to a more nominal climb setting. If that sounds involved, it becomes more so out of airports that are surrounded by high terrain, such as Butte, Montana. The southbound Butte departure involves a modified climb profile, where 20-degree flaps are retained until the airplane is much higher, and a course reversal in the climb, using 15 degrees of bank which keeps the airplane over the airport area until it reaches an altitude where a potential engine failure would be less of a problem.

Fig. 5-8. *You might consider back-taxiing the runway, instead, at uncontrolled airports; thereby assuring the runway is clear.*

Climb profiles such as this are developed with much careful study of the airplane's climb performance characteristics on one engine. The profiles plan for an engine failure, then write the procedure such that the airplane can maintain terrain separation while climbing IFR, until reaching altitudes where normal instrument procedures apply (Fig. 5-10).

If you fly a multiengine airplane, you would do well to consider its single-engine climb performance in selecting your initial flight path on a night departure, planning for the potential failure and your subsequent need to continue a climb without hitting any unseen obstacles nearby. If you are flying single-engine, select a procedure and flight path that would take you over terrain that you might possibly use for an emergency runway. Your best choice could easily be the airport runway you just departed, which would be possible if your flight path in the climb positioned your airplane over or near the field.

Instrument Departure Procedures

A good place to look for guidance in planning your night departure is the IFR departure procedures for your airport. Although many airports do not have IFR departure procedures, most airports with much night traffic do. A few dollars will get you some National Oceanic Service (NOS) charts for your airport, and they will outline a suggested route for IFR traffic. The route will often specify a minimum climb gradient and flight path to follow, assuming that outside references are invisible. It will even suggest minimum alti-

Fig. 5-9. *Those mountains on the horizon are invisible in darkness.*

tudes before continuing on course, based upon your direction of flight. As a night VFR pilot, you are certainly not bound to follow these IFR-specific procedures, but they may offer excellent help in avoiding obstacles in the area while you climb.

You may wish to create your own departure procedure. If so, the IFR departure might be a good place to start, then use a VFR chart, identify well-lit areas that should offer good outside reference and plan something that would allow you to climb safely and conveniently to altitude.

Take-off Performance

Col. Oliver North was questioned at a Senate subcommittee hearing about night departures from mountain airstrips in Central America. Since the heavily loaded C-141s and C-130s involved waited until nightfall to take off, surely they were hiding something. North replied that the performance advantages of a night departure enabled the aircraft to carry much heavier loads out of the tropical runways—a large enough difference as to render a daylight departure impractical. I thought the answer showed Colonel North to be quick on his feet. In reality, however, the practical aspects of a tropical night departure from a performance standpoint are slight—especially when considering the safety hazards of night operations out of unlit mountain airstrips. A night departure from a desert would certainly offer a huge performance advantage, perhaps doubling the airplane's payload, but probably not in the tropics. The humidity levels in tropical areas do much to

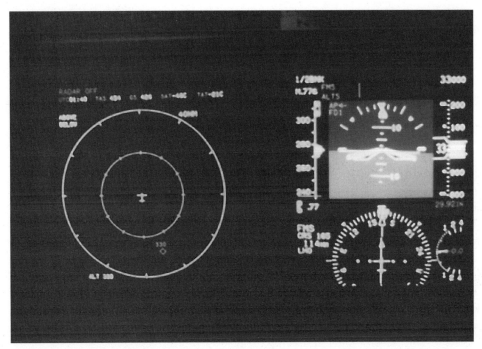

Fig. 5-10. *Level at 33,000 ft. The pilots of this Regional Jet carefully followed a climb profile to get there.*

moderate the nighttime changes in temperature. Where dry desert air might drop to near freezing levels in the night and greatly improve airplane performance, wet tropical air might drop less than 10 to 15 degrees. Whereas the night departures from Costa Rica certainly offer a performance advantage, it would be slight when compared to a nighttime departure in a drier climate. Although he spoke the truth about the concept of performance increases at night, the reality of the matter suggests that Colonel North was, perhaps, misleading.

There is more to take-off performance than just cooler air. Accurate control of the airplane's pitch attitude is critical on departures where maximum performance is desired. The lack of a visible horizon could make it very difficult for a pilot to determine pitch attitude, thereby causing some performance degradation, from a piloting skills perspective, at night, perhaps negating some of the advantage offered by the night air.

SKILLS TO PRACTICE

Night-time VFR departures place the pilot in a middle ground between VFR and IFR procedures. The ability to switch from visuals to instruments and back to visuals is critical for the night pilot. The view outside and the instruments must be constantly cross-checked; when they disagree, outside references become misleading, and the instrument indications must be judged to stand true. With this in mind a pilot considering a difficult night flight should be, if not instrument-rated, at least capable of flying for a while on instruments alone. The following exercises—variations of unusual attitude recoveries—are designed to improve the pilots skills in transitioning from external to internal references and vice-versa.

Unusual Attitude

You'll need the assistance of a competent pilot. At altitude, use visual references to place the airplane in a 45 to 60-degree banked, level turn. Trim the plane so that it will fly hands-off in the turn. Fly the turn for at least two circuits so that your internal balance equipment will become accustomed to the bank. While looking outside, put on an instrument training hood and transition to instruments. While flying by reference to instruments, reverse the direction of the turn, maintaining a constant altitude. After you have the airplane trimmed and stable, turning the other direction, remove the hood and reverse the turn again by visual references.

With the hood on and looking at the floor, have your partner slowly place the airplane in an attitude that is substantially different from where you started, such as a steep climb, turn or spiral. When your partner is ready, take the controls and stabilize the airplane in level flight using instrument references only. Repeat the maneuver using external references only. Practice the maneuver several times, recovering by both references, to a stable climb attitude.

Vertigo/Recovery

With the hood on and looking at the floor, have your partner cover all the instruments except the attitude indicator. Have your partner fly for a long time, doing so in a smooth manner, maneuvering the airplane through turns of at least two rotations each direction, then place the airplane into an unusual attitude. At that time, assume control of the airplane and recover by reference to the attitude indicator alone.

FURTHER READING

Lewis Bjork, "Digging Out of the Hole," *IFR,* October 1996.
R. D. Gless, *Night Flying,* AOPA Air Safety Foundation, vol. 16, pp. 1–9.
Lewis Bjork, *Piloting For Maximum Performance,* McGraw-Hill, New York, 1996.

6
Enroute navigation, maneuvering, and weather

Craig bent forward in the seat, shielding the dim lights of the panel with his hands. He put his face above the glare-shield, close to the window, and peered out into the darkness. Blurry lines of snowflakes raced into view, like incoming tracers, glaring in the landing light and fading the highway below. He switched the light off and the snowfall disappeared. Below, headlights from eastbound cars reflected off the highway, looking fuzzy, indicating that precipitation reached the ground. The obscurely red taillights of westbound cars showed the direction of his flight—the highway cut an almost straight line along the lake shore, piercing the desert—but he couldn't see far ahead. The dark world below faded just a few miles in front of the plane, crudely erased by an invisible overcast. For Craig and his three passengers, the sky was a black reflection of the ground below, lit only by the intermittent, sometimes blurry lights of cars, and they seemed to fade quickly a short distance away.

He knew the clouds were there, even though they couldn't be seen—that's the point. In better weather, the city would have been visible several minutes ago, the highway pointing at it from the mountains, laser straight from 70 mi away. Perhaps things would improve further down. He'd planned for 8500 ft but began a descent—perhaps to 6500 ft—to look for better weather. Craig did not want to lose sight of the road—that's the limit, the deciding factor, he thought. If he can see the road, he'll continue to press on; if not, he'd turn around. There was nothing particularly pressing about the flight anyway. They planned to eat dinner in Nevada after a short flight, returning home about midnight. If the weather worsened out there, they'd get some rooms at the hotel and make a little vacation out of it.

He'd tackled worse weather before. His son was instrument-rated—a flight instructor—and they'd flown together, in and out of clouds, all over the west. Craig certainly felt confident of his preparation for a flight like this, and he had plenty of equipment for the trip—more than enough. He loved the plane. A late model, turbocharged Piper Lance, it suited his needs nicely, and, when he felt up to it, he'd maybe buy a Cessna 340, or perhaps a 421. But for now, he had wings, enough power, and plenty of instruments to tackle a little weather.

"Piper _____, Salt Lake approach, I'm painting level 1 and 2 radar echoes along your route of flight, and lower ceilings reported to the west, please advise."

"Roger, ah, we're still VFR, request descent to 6500 ft."

"Piper _____, Salt Lake approach, altitude at your discretion."

"Roger."

Craig began a gentle descent. His wife chatted happily with their friends in the back. He loved this—flying around in the plane with friends, being a pilot—he'd always wanted to fly. He'd been a quick study and spared no expense in learning his pilot skills. There was some difficulty in getting a medical, which delayed his first solo a couple of years, but he'd kept right on flying. After 40 hours of instruction for his private license, he moved right on to spins, complex airplanes and instruments, still waiting for his medical application to clear. He'd received almost 70 hours of dual before finally soloing for the first time. A few cross-country flights and a checkride and he was finally a pilot—and a pretty good one.

There were a few ridges left to cross before the land flattened completely. The highway passed the lake shore, then bent north a few miles, around a short mountain range, then south around another, all in the space of 20 miles or so. To fly straight when the road bent would mean crossing a ridge—an uncomfortable thought at 6500 ft, with little or no forward visibility. As he looked for signs of the mountains, there was nothing at all to see. Night and weather blackened everything, except for the cars on the highway and a few sparse dwellings. He had to reference the gauges frequently, checking altitude and attitude—practically an instrument flight. He liked the challenge of instruments but didn't want to file IFR. He wasn't officially rated, and the clouds probably had a lot of ice. He decided to stick to the road.

It was cold outside, typical for the season—December, cold and wet—snows at night, with rain and sleet in the daytime. They'd departed in decent weather under the stars, but the sky clouded over just west of the shoreline. The weather briefing predicted almost clear skies and VFR along the entire route—so much for a forecast....

"Piper_____, Salt Lake approach, leaving the Salt Lake class bravo airspace, radar service terminated, squawk 1200, frequency change approved."

"Roger, g'night."

Outside, the road appeared to bend right, marking its passage around the first mountain ridge. He banked the plane to follow, but had to look hard to see anything ahead. The lights on the road dimmed rather suddenly, as the clouds seemed to reach lower, near the mountains. He lowered the nose, already passing through 6800 ft, trying to keep the highway in sight. The cars were much closer, and he could clearly make out a few road signs in the passing headlights. A few moments later, he followed the road as it turned back to the south, the plane flying a few hundred feet above. Both the plane and road descended slowly. He felt relieved to have crossed the ridge, hoping the ceilings beyond would raise up a bit. He was flying uncomfortably low in order to see the road, and he could *feel* the unseen mountains nearby. *Can't afford to lose the road—and can't go much lower.*

About then, night swallowed the plane—the clouds simply reached down and ate it. The highway disappeared, and rain and snow beat against the windshield. They were solid IFR and Craig could barely see it coming. The little piece of highway that reached ahead of the plane seemed to retract, inwards, into blackness, and then the rain—they had to be picking up ice. It was impossible to go any further; and equally impossible to stop.

He began a left turn, to reverse course. Completely IFR, he noted the heading, set the bank angle, and adjusted the pitch attitude to continue the descent, hoping to break out of clouds early in the turn. He had to get a visual reference soon, no telling exactly where the mountains were....But then his world changed again—instantly.

With a severe jolt, the engine crashed its way through the firewall, the aft cabin crumpled, passenger seats broke from their tracks and the four bodies inside crushed as the airplane hit solid, impenetrable rock. The little Piper stopped dead, flattened like a bug swatted with a granite hand. The surprise and shock of the impact took a moment to register as noise abruptly ceased, leaving only a deafening silence touched by lightly falling snow. His body broken and unable to breathe, Craig felt little...then nothing at all.

Accident occurred DEC 18, 1994 at Grantsville, Utah.

Aircraft: Piper PA-32RT-300

Injuries: 4 Fatal.

The non-instrument rated private pilot received a weather briefing for a VFR flight over 5 hours before he actually departed. He and his three passengers departed at night in mountainous terrain and in VFR conditions with the intention of flying to an airport located 90 miles away for dinner. The pilot received ATC radar advisories and reported that the ceilings were getting lower along his route of flight. He was advised by ATC that areas of level one and two precipitation existed in front of him. The airplane continued to descend after ATC services were terminated. Radar data for the airplane was lost shortly thereafter. The airplane impacted a mountain ridge about 6,200 ft MSL and was destroyed. The ridge is located along a direct line from the departure airport to the destination airport. No distress calls were recorded from the pilot, and no evidence of pre-impact mechanical deficiencies were found. Localized adverse weather conditions, including low ceilings and snow, were reported moving west to east as the airplane flew east to west.

Probable Cause

The VFR pilot's attempt to continue the flight into instrument meteorological conditions, and his failure to maintain altitude/clearance with mountainous terrain.

BACKGROUND INFORMATION

More than 10 percent of night accidents involve continued VFR flight into IFR conditions/weather (Fig. 6-1). Just like Craig, pilots attempt to remain VFR in lousy weather, avoiding the invisible, and often crash after maneuvering at low altitudes. Another 25 percent of the total involve weather in some way, and 28 percent happened while the plane was en route. According to these figures, the odds of crashing because of weather or en route phenomena are much higher than the dreaded engine failure. With this in mind, consider the following points with the same care as you would the sound of your engine in the dark.

Visibility

You can see stars at night, from billions of miles away, and not at all in the daylight (Fig. 6-2). The stars make their own light, and when your eyes adjust to the darkness of

Fig. 6-1. *The majority of night accidents involve weather in some way.*

night, they twinkle in all their glory. Dim lights on the ground carry the same feature. By day, a town might be nothing more that a few scattered buildings, you'd have to be nearly on top of a small one to pick it out from any distance. But at night, when things are really dark, and the air is clear, you can see a small town traffic light from 2 mi high, pick out a solitary airport beacon from 100 mi away. However, the landmarks you customarily use for navigation and pilotage in daylight may very well be invisible at night.

Daylight pilots often make use of railroads, power lines, mountains, rivers, and lakes for navigation (Fig. 6-3a). Few of these identifying features would be visible on a dark, moonless night. It is for this reason that night navigation requires landmarks that are self-lit. This is easier than it sounds, because the most obvious landmarks are cities and highways—and you're probably flying near or to one of those cities anyway (Fig. 6-3b). Combine city lights with great visibility, and the VFR night pilot can easily pick out city landmarks from distances that would be unheard of in daylight.

Seeing and Avoiding Weather

In the same way that cities are visible from great distances in darkness, thunderstorms, by virtue of their lightning, are even more visible. I have flown in areas of thunderstorms where towering cumulus could be evidenced in darkness far beyond the actual horizon. Many storms reach such heights that their lightning flashes can be seen from the air for

Fig. 6-2. *You can see lighted objects (like the moon and stars) from incredible distances, at night.*

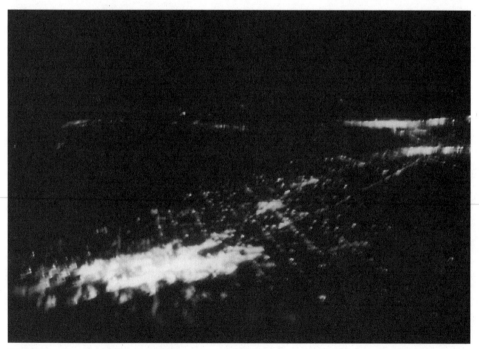

Fig. 6-3. *The navigational references used at night are often different from those used in daylight.*

hundreds of miles. Despite the spectacular visibility of thunderstorms and lightning, most clouds just add to the nightly darkness. Looking for them is like trying to find a shadow in a dark room—without light of some kind it's an impossible task. A pilot might deduce an overcast because of the lack of the expected stars, or even notice the stars "blinking out" as the airplane passes under a cloud. Passing over a cloud can be more subtle, however, as the pilot has to pay close attention to the ground lights and notice the changes as clouds blot them out. If the lights below are bright, as over a large metropolitan area, lower clouds will often be highly visible, appearing to glow in the city lights.

Clouds, because of their whiteness, handily reflect moonlight if the airplane is properly positioned. Moon-lit clouds are much easier to see from a position above than from below a thick overcast—where the moonlight might not penetrate (Fig. 6-4). Generally speaking, though, bright moonlight conditions may greatly ease a pilot's task in avoiding clouds and weather. In the absence of the moon, however, when the night is really dark, actually seeing clouds will depend on the area over which the plane is flown and the type of weather involved.

Diurnal Temperature Shift

The first weather change that occurs when the sun sets is a general cooling of temperature. As the earth rotates, half of its surface is bathed in sunlight and the other half in darkness. There is a daily shift in temperature from night to day, referred to as a *diurnal*

Fig. 6-4. *Clouds like these are visible under a bright moon.*

125

temperature shift. It is this constant adjustment of temperature that fuels much of the world's weather—the basic premise being that weather is caused by differential heating of the earth's atmosphere. The peculiarities of night-related weather are basically tied to this one simple precept—that it gets a little cooler during the night, warmer in the day.

An interesting feature of this daily temperature change is that sunlight itself does not directly heat the air—it passes through it. The sunlight heats the surface of the earth, and warmed surface emits heat energy into the atmosphere, in a process called *radiation,* which raises the temperature of the air above. Air closest to the surface of the earth is far more affected by surface heat radiation than air at high altitudes. The Space Shuttle demonstrates this process on a smaller scale, when it has just landed after a hot reentry. If you stood near it, you would feel the heat energy radiating from its frame—which is why many of its recovery crew wear heat-reflective clothing. After a time, the ship cools to ambient temperature and can be approached safely—it has radiated its heat away. As the surface of the earth becomes warm in sunlight, this warmth is radiated into the atmosphere above and elevates its temperature. When the source of heat—the sun—is taken away, the earth continues to radiate its heat during the night, slowly losing energy until the sun rises sufficiently to be an effective heater again. Because of this process air temperatures will generally cool throughout the night, reaching their lowest point sometime a little after dawn.

The topography of the earth's surface may greatly affect the temperatures of the air above. In the poles, where sunlight is indirect, little radiative heating takes place, because the surface itself is not particularly hot. It is interesting to note that in the absence of heat from radiation, the air seems to assume a uniform temperature of about $-56°C$. This temperature is an average found at the coldest regions of the poles as well as at very high altitudes worldwide. It's about as cold as the air is willing to go and, away from the effects of surface radiation, remains fairly constant in a diurnal cycle. The atmosphere high above you, even in the summertime, will reach these low subzero temperatures. The air at altitude is far enough separated from the earth's radiative heat as to be virtually unaffected by the warmth near the ground. For this reason, most nightly weather changes occur at lower altitudes.

Humidity-Related Temperature Changes

The atmosphere has the ability to contain, in suspension, all sorts of foreign substances, floating them about like clouds, for days. Environmentalists are delighted to educate the public on the hazards of smoke pollution, exhaust, ozone, and various chemicals, perhaps taking advantage by imposing regulatory changes. Although all suspended particles may affect the atmosphere in some way, the airborne substance most relevant to this discussion is the constant presence of water, sometimes huge amounts of it, in the air we breathe and fly in. Air's ability to contain water is directly related to its temperature. The warmer the air's temperature, the more water it can contain. The air at the earth's poles is among the driest in the world, for example. Travelers through those regions commonly suffer from dry skin maladies. Perhaps you've heard of the concept of freeze-drying. This is a preservation technique which removes water from organic materials by freezing and sublimation. Warm air, however, has a high capacity to retain water, as evidenced by conditions in the tropics. Humidity is expressed in relative terms, thusly: *Relative humidity* is the ratio of the existing amount of water vapor in the air at a given temperature to the maximum amount that could exist at that temperature; humidity is usually expressed in percentages.

Temperature is directly related to humidity and is part of its definition. In fact, humidity can also be expressed in terms of temperature. *Dewpoint* is the temperature to which the air would have to be cooled such that it would be saturated (i.e., 100 percent of capacity) with its existing water content. If the air is cooled to a temperature below its dewpoint, the water vapor it contains would condense and become visible liquid, whether as clouds, frost, or precipitation. One definition states it like this: *Dew point* (or dew-point temperature) is the temperature to which a sample of air must be cooled, while the mixing ratio and barometric pressure remain constant, to attain saturation with respect to water.

Moisture present in the air affects the rate at which temperature changes may occur. Water possesses a chemical property referred to as a *high specific heat,* meaning that it requires enormous amounts of heat energy to change the temperature of the water. You can personally experience this feature as you increase the heat of a cold swimming pool; it may require a very long time to warm up appreciably—you might postpone your swimming for a couple of days even. This phenomenon is also evident in a comparison between the diurnal temperatures of arid and moist climates. Dry desert air might cool to near-freezing temperatures at night and become dangerously hot in the daytime. Conversely, the moist air of the tropics demonstrates characteristics indicative of a high water content, with temperatures changing relatively few degrees from night to day.

Small air masses can become affected by water content simply by virtue of their position. Air over the ocean will typically be wetter or more humid than air over dry land. The ocean itself radiates heat into the atmosphere slower than dry land, which causes the air above water to heat more slowly, and at night moist air above an ocean will cool more slowly, thus becoming slightly warmer than its over-land counterparts. In summary, land surfaces heat and cool faster than water surfaces, and the air above each is correspondingly affected.

Land and Water Phenomena

The difference in temperature characteristics of air over land can give rise to wind in the same way as air over water. During the day in coastal regions, warm over-land air begins to rise. The rising air creates a low pressure area over land. Relatively moist, cooler air from the sea is drawn in by the low pressure, causing a breeze, that blows inland—a sea breeze. At night, as temperatures over land cool more rapidly than those over the water, the circulation reverses (see Fig. 6-5). Now the air over the sea is relatively warm, compared to its counterpart over land. The ocean air rises, creating a low pressure which draws the inland air out to sea, creating a land breeze.

A similar process can create wind in mountainous areas. The upper slopes of a mountain in sunlight may radiate heat, warming air at higher altitudes. The air at equivalent altitudes

Day

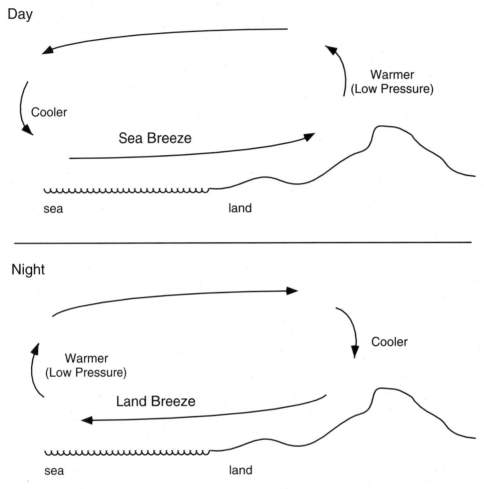

Warmer
(Low Pressure)

Cooler

Sea Breeze

sea   land

Night

Cooler

Warmer
(Low Pressure)

Land Breeze

sea   land

Fig. 6-5. *Winds like this may happen only in the absence of more powerful weather systems.*

over a nearby valley is relatively cool, being somewhat distant from ground and its radiant heat. Accordingly, due to the relative temperature differences, a low pressure forms near the higher elevations of the slope. The resultant wind blows up the mountain slope, emanating, as it were, from the valley. It is commonly called a *valley wind.*

At night, the air in contact with the mountain slope cools relatively fast, due to its proximity to the land surface. The cool air descends the mountain slope, toward the relatively warm air below. The daytime circulation effectively reverses, and a *canyon wind* blows (Fig. 6-6).

For temperature-related wind phenomena to occur, the overall wind condition, or pressure gradient from existing weather, must be weak. When winds blow for other reasons, such as frontal activity, the localized topographic winds become mixed into the larger picture and tend to follow the flows of more powerful weather systems.

Day

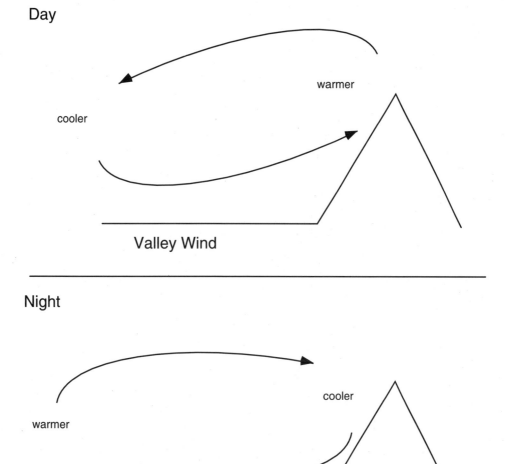

warmer

cooler

Valley Wind

Night

cooler

warmer

Canyon Wind

Fig. 6-6. *Along the eastern slope of the Rocky Mountains, canyon winds are known to be amazingly powerful—to the point of causing damage.*

Diurnal Fog and Clouds

Nighttime may produce conditions ideal for the formation of fog. If a moist air mass is stationed over land at nightfall, the land cools quickly and cools the air in contact with it. The cooling air soon reaches its dewpoint, and condensation occurs, forming fog. Radiation fog is usually rather shallow but quite opaque laterally. To a pilot flying above, the ground surface directly below might be visible, and tall objects such as airport control towers may actually be seen above the fog (Fig. 6-7), but for the most part, the fog obliterates a useful view of the ground. If the temperature-dewpoint spread is close at sundown and there is little wind, you might expect some radiation fog to form during the night.

With the help of a friend, I packed several tools into a Mooney and we departed early one morning. The sunrise reflected brightly from snow-capped peaks under a brilliant blue, clear sky. My friend asked if I checked the weather. "We don' need no stinking weather," I said, with a dash of bravado. "Visibility is a hundred miles, and we're only flying 60 miles away—no clouds in sight." The Mooney covered the distance in 20 minutes. From overhead we could see the valley where the town lay, but no town in sight. It was buried in a layer of radiation fog. The air over the valley was moist to begin with and the wind perfectly still. During the night, the land cooled the air in contact with it and the air condensed. We couldn't see much, but it was apparent that the fog was not particularly thick. Having been there many times, I knew exactly where the airport was and,

Fig. 6-7. *The Wind River mountain range from 33,000 ft in the early morning. Low clouds and fog, like this, are often worse at night.*

amazingly, found the tower sticking above the fog. We circled for 45 minutes before the sun burned off enough of the fog that we could land. My friend chides me to this day, mimicking my best accent, "We don' need no stinking weather."

Other types of fog are caused by *advection,* when relatively warm, moist air from over a water surface is blown across cooling land during the night. The land cools the air to its dewpoint, and fog appears. Advection fog can be quite thick, even forming a low stratus layer, depending upon the strength of the wind. This is the mechanism behind the famous fog banks in London, San Francisco, and Los Angeles. Because of advection fog, a pilot departing at sunset toward the coast might easily find airports completely socked in by morning, and have to wait many hours after sunrise for the conditions to change.

Weather

All this differential heating at the surface of the earth can generate many kinds of weather. The simplest is possibly a thermal (Fig. 6-8). Suppose we look at an asphalt parking lot in the middle of a cornfield. Under direct sunlight, the parking lot becomes a hellish place, reminiscent of many airports, where you would not readily walk barefoot. The place heats the air immediately above it. Since the air above the cornfield is relatively cool, the air over the parking lot begins to rise. It may occur as a bubble, like what forms in a pan when water boils, or a column of continuously rising air; either way, the region of rising air over the parking lot is called a *thermal* and is by its nature associated with hot air. A pilot flying across the cornfield would feel a sharp jolt, or turbulent bump as the plane passes over the parking lot. A glider pilot, forever looking for rising air would take position over the parking lot and circle, hoping to make use of the invisible thermal to gain altitude.

As the warm air in the thermal gains altitude, it cools—it is moving further from the surface, which is the source of its heat. If the air is at all moist and cools sufficiently to reach

Day

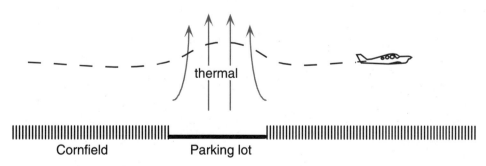

Fig. 6-8. *Thermals are caused by differential heating at the surface.*

its dewpoint, you'd see a cloud form. Invariably, clouds formed by thermal activity are cumulus (Fig. 6-9). Once the vapor in the cloud condenses to water, the thermal can be seen. Glider pilots treat cumulus clouds as aerial signposts, indicating the presence of thermal lift. If the air is sufficiently moist to retain its heat energy to high altitude, the cloud itself may become rather powerful and perhaps make a little rain, like a thunderstorm.

If we go back to the humble beginning in the parking lot, however, we see a reversal of trends after the sun sets. Without sunlight, the asphalt quickly cools, perhaps even more rapidly than the surrounding cornfield. The cool air over the parking lot descends and, upon reaching the ground, spreads outward into the surrounding warmth of the corn. The hapless glider pilot towed aloft too late in the day will find not lift over the parking lot but dreadful sink. The building clouds above will also experience a reversal of trends, eventually shrinking back to earth and fading away completely. In cases of powerful storms, the lift they once generated might reverse into rather powerful sink during the cool of the evening and a flight below them might encounter a serious case of windshear which would diminish rapidly as the night wears on. Eventually, the bumps fade away completely in the quiet night—you remember, they were caused by heat in the first place—until the pilot flying over the parking lot well after dark will be pleased by a smooth ride (Fig. 6-10). Such is the diurnal rise and fall of convective activity.

Icing

Cool night conditions may produce an icing hazard. For ice to form, airborne water must be in a liquid state. You have to be able to see it, such as in clouds and precipitation. Humid conditions, where the water is gaseous and invisible, will not produce structural icing. As the night air cools, however, you remember that the air's capacity to retain water diminishes. Eventually the air will become saturated with existing moisture, and any excess will condense to liquid form. You will see clouds form. As conditions become

Afternoon

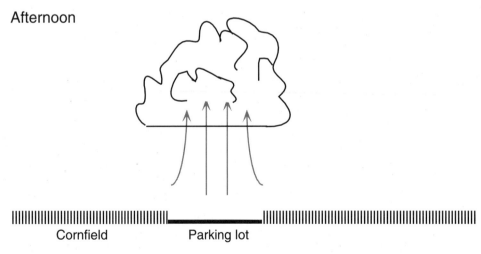

Cornfield   Parking lot

Fig. 6-9. *Sometimes a strong thermal will form cumulus clouds.*

Late evening

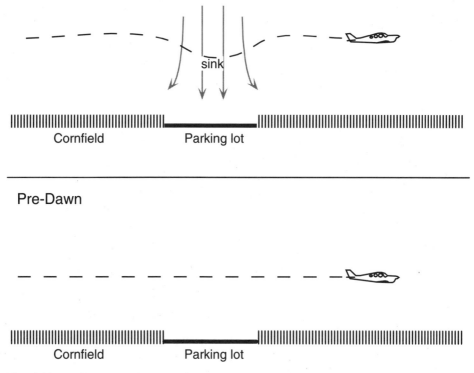

Pre-Dawn

Fig. 6-10. *Without sunlight, most thermals fade to glass.*

cooler, the clouds may form at lower altitudes. Any time a plane flies though clouds (which are visible moisture) in near-freezing conditions, there is the potential of structural icing. Night enhances that hazard both by cooling the air and by increasing the humidity levels nearer to the point where the potential for clouds and icing increase.

In very cold temperatures, visible water may be already frozen. The air itself may be too cold to retain much moisture, and that which is visible is already ice. Flying through clouds of this type usually presents little icing hazard. Water that is already frozen will not stick to your plane.

Your airplane's carburetor has the ability to create icing conditions internally, while the rest of your airplane could be flying in clear air. Most carburetors make use of a venturi of sorts, to produce suction required for proper mixing of fuel and air. As air rushes at increased velocity through the venturi of your carburetor's throat, its temperature decreases to colder than outside air. This temperature decrease, in humid conditions, might be sufficient to cause liquid condensation within the carburetor while freezing the liquid as it forms. The conditions may occur when the ambient air is well above freezing, perhaps as high as 60°F. Ice forming in the carburetor could kill your engine. Carburetor heat usually diverts warm air into the carburetor from elsewhere in the engine or heater compartment, which one hopes will raise the temperature of the air in the carburetor to above condensation/freezing levels and eliminate the ice.

Some operational procedures suggest the application of carburetor heat whenever a throttle reduction is considered, such as prior to landing. The specific procedure completely depends on the specific manufacturer and airplane type and is often the subject of some debate. I know of one crash, for example, which happened because ice formed in the carburetor with the heat on. On a very cold day in January, an instructor and student practiced slow flight in a Cessna 150. Carburetor heat was applied, as per Cessna procedure, when the student throttled the engine back and slowed down. With full flaps applied and the airplane sufficiently slow, the student applied throttle to stabilize the airplane at altitude, and the engine promptly quit. The pilots survived the crash-landing in a field, but the airplane did not.

What happened? Normal temperatures in the carburetor on such a cold day were well below freezing. No ice could stick in the carburetor because the low temperatures would not allow water to condense as a liquid. When the student applied carburetor heat, the temperatures within the carburetor were increased to a point where ice could form nicely. The intake throat began to ice up and restrict the air coming into the engine, but it did not yet quit, because the air supplied at the low power setting was still adequate to sustain combustion. When the student advanced the throttle, however, the increase in fuel flow could not be appropriately matched with air (the intake was restricted by ice), and the engine quit because of the rich mixture. The engine essentially choked to death. In retrospect, carburetor heat should never have been applied in such cold conditions—it wasn't needed anyway. When the engine did quit, it could possibly have been made to run again by leaning the mixture, although it most likely would not have developed full power.

Pilotage

Night VFR navigation is highly dependent on what is available to see. If the plane is flown over populated areas, following roads or aiming directly at cities, which are sometimes visible from long distances, is a simple matter. As ground lights become more sparse in the more rural areas, however, the pilot has to pick the best of unlighted landmarks. Water is sometimes a good choice where there is enough ambient light to produce a reflection. Stars can provide a quick sense of direction, but you must not forget that stars move. Navigation by the stars on a long flight could inadvertently take a pilot many degrees off the intended course. You might best use stars when you can cross-check your heading with compass and the occasional landmark on the ground. Roads can be handy, especially if they are well traveled. Cars on a highway are easy to follow, and chasing roads and rivers will generally help a pilot avoid high terrain (Fig. 6-11).

Fig. 6-11. *Large rivers can be handy navigational references in darkness.*

IFR

Instrument procedures are often in conflict with normal VFR navigation, and you may be hard pressed to fly under visual rules and still maintain a position where a quick transition to IFR is possible. Instrument flight rules, for example, require the airplane to be flown at an altitude that guarantees obstacle clearance anywhere along the route. Instead of going directly from city to city, or following roads, an instrument course often involves small kinks and jogs in the flight path to track ground-based navigation aids.

In Craig's case, he chose to follow a road under a low overcast, descending well below the minimum safe altitude for instrument flight. The weather included embedded storms and the strong potential for structural icing. By electing to remain VFR he avoided the weather but of necessity in following the highway placed the airplane too low for a safe transition to IFR flight. You'll need to consider the trade-offs between the two. If you plan to go IFR, you'd best have equipment to tackle the weather because you'll be flying through it. If you remain VFR, you'd best ensure that existing weather conditions are adequate for you to stay that way.

Dead Reckoning

It's possible to navigate in visual conditions for a surprisingly long way on your compass alone. Lindbergh basically aimed at the letter *E* on his compass for 30 hours over the ocean. Admittedly, Europe is a pretty large target to aim at, but in this case, the proverbial dead reckoning form of navigation was entirely adequate. Flying at night in the absence of good landmarks, dead reckoning is a completely viable solution. You might be able to obtain position fixes with electronic navigation equipment and simply fly the proper compass heading for great distances.

Traffic Avoidance

Finding legal traffic in dark visual conditions is a delight. When a lighted airplane is framed against a dark sky, it can be visible from incredible distance—just try spotting it from 100 mi away in the daytime! However, when flying above a city, traffic below your altitude could easily blend into the city lights and become practically invisible. If you expect to notice its movement, which, by the way, is about the only way you will spot traffic below, keep in mind that many of the city lights below are cars—and they move, too.

Illegal traffic, i.e., airplanes without lighted position lights, are invisible at night. Cities of strategic importance learned this by heart as bombers pounded them during nighttime raids in World War II. Shooting many bombers down required a great deal of luck and powerful search lights. Likewise, smugglers and pilots who simply forget to turn their position lights on are impossible to see in darkness.

In general, however, it is far easier to spot other airborne traffic in darkness than in daylight (see Fig. 6-12).

Fig. 6-12. *At night, this airplane's lights would be visible ahead of its contrail.*

TECHNIQUE

A cross-country flight in darkness demands numerous decisions of the pilot which must be based on outside sources of information to a greater degree than in daylight. Weather, for example, is often unseen in darkness. The pilot's weather decisions, and subsequent choices of route, navigation methods, equipment, and fuel, are generally based upon the professional assessments of outside sources. It is vital that the information upon which a pilot bases such decisions is current and relevant to the intended course. In Craig's case, his weather information was 5 hours old. He felt that his information was adequate, based upon the VFR assessment he received earlier. It was unexpected changes in the weather that led to his demise. He assumed that the weather was adequate and, upon encountering adverse conditions, hoped that it would improve further on, no doubt based upon the information he received earlier.

If he had flown through the same conditions in daylight, his own assessment of the weather as seen out the windows would have probably convinced him to turn around much sooner. At night, the weather ahead remained cloaked in mystery, his only indications of its existence were a vague radar advisory and his fading view of the lights below. In retrospect, pilots encountering unexpected weather in darkness, with indications such as these, should assume the worst and divert to better conditions. It is impossible to navigate through and around dangerous weather by Braille—a pilot has to see (Fig. 6-13). The type of weather flying Craig attempted is often referred to as *scud-running* and should be avoided after dark, particularly in mountainous terrain.

A pilot must consider several factors involving weather and darkness. When weather is forecast or reported to be near to the route, in darkness, the airplane must be prepared to fly *through* the weather as described, since in practice the pilot will probably not be aware of the weather until the airplane has actually penetrated it. With this in mind, the plane must have adequate equipment, perhaps to include deice and anti-ice systems and storm detection equipment, not to mention the instruments needed for comfortable IFR flight. If the flight is restricted to VFR by virtue of equipment limitations or those of a rating, the pilot must be especially sure of the weather's location and modify the route to stay well clear or take the risk of inadvertent IFR. My personal limitations when considering a night VFR cross-country flight are CAVU—ceiling and visibility unlimited. If it is otherwise, I will plan to fly IFR, and prepare accordingly.

If unexpected weather is encountered in flight, the VFR pilot must take immediate action to avoid it, which might often involve a course reversal. If the flight is IFR, the pilot must immediately ascertain the conditions and take appropriate protective measures, such as activating radar or perhaps avoiding icing conditions. In both cases, the pilot should report the type and location of the weather to ground facilities and obtain a radio update of conditions. For the VFR pilot, communication with a source of current weather could play a critical role in the pilot's decision making with regard to a course reversal, diversion, or alternate airport.

If unexpected weather is encountered on a VFR flight plan, and the pilot opts to transition to IFR, the legal procedures in controlled airspace require the flight to remain in VFR conditions until an IFR clearance is received. Maintaining VFR conditions in the

Fig. 6-13. *As the light fades, these clouds will become invisible, and so will the mountains behind them.*

vicinity of invisible weather can be a very difficult task. With this in mind, the process of transition to IFR is facilitated if the pilot is already in communication with ATC and perhaps located on an IFR suitable route and near an appropriate IFR altitude. In other words, it is better if a little planning has already occurred if the pilot suspects bad weather to be out there.

Decision Aids

Without the aid of visual information to establish the limits and definition of weather, a pilot might be greatly aided by drawing a picture of the weather directly on the navigation charts. Nothing fancy is required, here, just that the pilot block out areas of reported weather as depicted in a synopsis or weather pictorial. En route, the pilot could easily refer to this information and visualize the location of hazards and select course and destination accordingly.

Unlighted airports should be practically eliminated from the night en route decision process. Perhaps the chart should be further detailed so as to point out airports with adequate lighting. An unlighted airport should be considered for use after dark only as a last resort, in a condition bordering on the seriousness of an emergency landing, because without runway lights, an emergency could easily occur.

In areas containing large unlit hazards such as mountains, the chart might be further detailed to indicate preferred routes, or general areas to be avoided, perhaps to coincide with established instrument procedures.

A pictorial representation is the key to a quick decision. Night VFR pilots could benefit from the pictorial diagrams represented on instrument approach charts, as they highlight preferred routes around terminal areas and offer suggestions of areas to avoid. Likewise a good VFR chart is very handy to the IFR pilot in understanding the nature of the hazards below, perhaps assisting in the selection of alternate airports and fuel needs on the basis of the terrain surrounding a prospective airport.

SKILLS TO PRACTICE

The regulations require a pilot to become familiar with all pertinent information about departure, the route, and destination that would affect the intended flight. This includes weather, airport data, terrain, distance, fuel, and hundreds of other things, such as the phone number of the car rental company. In the interest of increasing your in-flight situation awareness, attempt to compile this information onto your chart, in pictorial format where possible, before you make your next cross-country flight (Fig. 6-14). Try to avoid relating loose papers, such as flight logs, AFDs, and verbal weather data while you fly. The more you can reduce the data to a useful format, the more useful and available it will be.

FURTHER READING

AC-00-6A, *Aviation Weather,* Department of Transportation, Federal Aviation Administration, Flight Standards Service, rev. 1975.

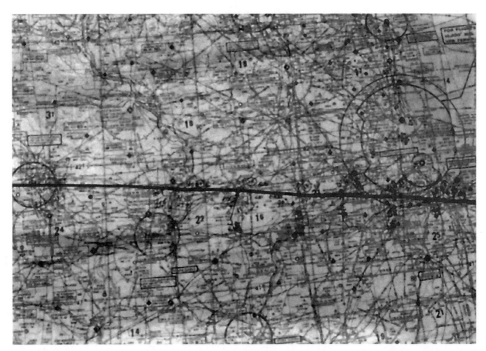

Fig. 6-14. *If weather is suspected, the pilot might draw it or mark its location on the chart.*

7
Approach and landing

Grant couldn't wait to get to the airport—he didn't want to wait any longer, anyway. The car in front made it necessary, sitting like an idiot in the face of a green light. Grant leaned on his horn and shouted, "Hey, buddy, it's the pedal on the right!" The driver ahead looked up suddenly and floored his gas-pedal, launching across the intersection. Grant moved ahead, but traffic had already built up at the next light, forcing him to wait in line again. It looked like rush hour, even though the hour had supposedly passed. Grant had gone home from work, changed clothes, grabbed his flight bag, kissed his wife, and dashed out the door. A recently certified Private Pilot, he headed for the wild blue yonder, a flight out of traffic and city mess. His frustration grew with each passing minute, the airport still a few miles west and the sun already going down.

To the east—125 mi behind him—waited the little airport at Manila. Grant's brother had left earlier with the pickup truck, loaded with camping gear, ammunition, and rifles. Since he could not get released from work early, Grant felt that he could easily rent a plane and fly out to Manila, where the deer hunt could start in earnest. The brother brought Grant's handheld radio in the truck. They picked a discreet communications frequency and planned to rendezvous at the airport a little before sunset, but Grant was running late, and the traffic didn't help.

He had begun to curse, muttering under his breath at every car that would even think of slowing him down. He's gotta fly, dang it—get outta the way! Sometime after parking the car and dashing into the FBO for the keys, he began to calm down. Flying at the airport was something to be savored, not rushed. He walked toward the little Piper Warrior, admiring its pretty lines, thinking about the flight ahead. He'd fly right over those miserable idiots in their cars.

He pulled the plane out of the shade hangar and did a preflight, delighted to be away from traffic, from work, from the cares of city life. He looked forward to camping in the Uintahs, eventually sighting a deer down the barrel of his rifle. Things were definitely looking up. The sun began to slip below the mountain ridge as he taxied to the active runway. The sky lit up with a blaze of gold, hardly a cloud in sight. It seemed fitting, for Grant was feeling pretty good. He departed into glassy-smooth air, turned eastward, and settled down to watch the scenery below, slowly fading into a lavender night. Mountains slipped by, their granite tops awash in amber light, reflecting pink, fading to purple in the lower darkness of the valleys. The light continued to dim as he flew eastward for a little over an hour. The high Uintah range loomed, barely discernible, off the right, with the autumn-yellow aspen groves on their slopes showing only a lighter shade of gray than the dark pine forests which spread out below.

Manila is not a big town, hardly more than a service station and a few mobile homes, but Grant had no trouble finding it. The highway slanted southeast from Evanston, wound through some badlands, and eventually ran into the lake. *Keep the Uintahs on the right, the highway will come in from the left, and if you cross the lake, you've gone too far.* He spotted the town just fine from 50 mi away. Getting closer, he could make out some lights, the highway—-and yes, the airport beacon. It was pretty dark down there.

He popped open his flight guide and looked up the airport. That must be the beacon, the runway is paved, runs east-west, and is 5300 ft long; elevation 6175; use 122.8 for advisories. He set 122.8 in the radio and keyed the mike several times while circling the airport. Expecting to see lights, all he saw was blackness, and a few cars on the road. Where's the dang runway? He keyed the mike and asked for an advisory, but no one responded—nobody home. He switched the radio frequency to 123.45—

"Hey, bro'. This is Grant, you listening?"

"Yeah. 'Been watching you circle for the last five minutes. What took you so long?"

"Traffic."

"Well, fine. Get your butt down here, I'm tired of waiting."

"Uh, do you see any runway lights down there?"

"Nothing lit, but I'll drive out on the ramp and have a look."

Grant watched as headlights flicked on below and changed to high-beam as a vehicle began to move about on the airport. As the truck maneuvered about down there, Grant could see details on the ground where its lights shone. He began to wonder....

"There's nothing but little reflectors down here. Don't see any lights except that big one that spins around on the pole."

"Shoot. Look, if you park the truck facing down the runway with your headlights on, I could probably use them to land."

"You sure about that?"

"Hey, I'm a trained pilot—this is nothing."

"Okay."

A few minutes later, the truck was situated at the end of the runway, headlights shining on high-beam into the darkness. Grant pondered the situation from above. The lights seemed brighter before. Out on the runway, they hardly reached past the numbers. He didn't like the idea of landing over an obstacle like a truck. It seemed kind of unforgiving, like approaching an aircraft carrier, or a cliff, or something. Come in a little too low, and bam! It's over. Probably better to land the other direction, into the lights. He adjusted the flight to make a pattern for landing and began the descent. He was trained to use a point abeam the approach numbers as the starting point for the approach, but the only numbers barely visible were in front of the truck, at the other end of the runway. He didn't want to land so far from the truck that the headlights would be useless, but to land too close and overshoot would be a disaster. What would the propeller do to a truck, anyway?

With a mile of unseen runway to work with, Grant picked a place comfortably far from the truck and throttled back to 1800 rpm. He allowed the airplane to slow down and pulled the flap lever to its first detent; 80 knots registered on the airspeed indicator with a 500 fpm descent. At this point, he looked back over his left shoulder, again looking for the approach numbers and the position from which he would turn base. He could see the truck headlights and had a vague idea of the runway, but nothing much else. He turned base simply because he thought it would be about right, not really knowing exactly where the runway began. *I should have asked about the wind,* he thought.

"Hey, what is the wind doing down there?"

"Blowing. What else?"

"No. Which direction is it blowing?"

"Uh, which direction am I facing? Let's see, it's coming from behind me, maybe a little from the left."

"Hard?"

"No. Just a breeze."

Since he was landing head-on to the truck, that would make it a right quartering headwind, but probably negligible because it wasn't blowing hard. There was no windsock visible, and other indicators like foliage or smoke were not visible either. *Okay, figure a light crosswind from the right.* Grant turned final, aiming directly into the headlights. His own landing light was on, and he concentrated intently on its beam, hoping to see some indication of the runway. He pulled on full flaps and slowed to 70 knots, adding a little power to slow the descent—but a descent into what? It felt uncomfortable. He was not very high and couldn't make out any details in the blackness below—hopefully the runway would come into view pretty soon. He added a little more power, being cautious with the descent, approaching the unknown like you would walking barefoot through your garage with the lights off. Grant felt butterflies in the pit of his stomach. What if he misjudged the runway's position and landed on a house? Continuing much further might invite a collision with a tree or something.

Finally, when he was about to pour on the power and go around, a field appeared below, edged with a long line of barbed-wire, with the airplane perfectly positioned to land on the fence-line. A little to the right was an irrigation ditch, and a couple of bushes—no tall trees so far. Grant's hand trembled slightly on the throttle, but he kept looking—a little further to the right and, happily, runway markings emerged, gray against the night. He corrected that way, banking first right, then quickly left, to settle over the center-line—relieved to have found the strip at last. He'd kept the speed up while rushing to adjust for the runway, and went into the flare with a little power and about 80 knots. As the nose came up, his gaze followed the beam of the landing light, looking further down the runway, and what he saw tied his stomach in a knot. The truck parked RIGHT THERE, unseen huge and solid in the darkness, behind high-beams blasting into Grant's face. At about the same time, while staring into the bright headlights, he lost all sight of the runway. The wheels had not yet contacted pavement as the plane wafted blindly toward a collision at 70 mi/hr.

Grant had to get the plane on the ground and stopped immediately, and he couldn't see to do it, so he pushed the yoke forward and chopped the power. Staring wild-eyed into the blazing lights, his actions were panicky, like a deer caught in the brights of a car. His feet locked up on the brakes and he felt the plane lurch, hopping and, skidding, down the pavement. The screech of the tires quickly changed to a loud rumble, and pieces of twigs and grass splattered up into the windshield, momentarily hanging in the air like insects, illuminated from behind, before spattering onto the plane. Seconds later, a powerful jolt threw him forward and right against the shoulder harness, there were several loud thumps which shook the plane and a sound much like metal garbage cans being run over by a truck, and then all noise abruptly hushed as the plane came to a halt.

The instrument lights still glowed softly, and the whine of gyros could be heard against a background of near-silence. Grant was shaking. He could see dirt and sagebrush all over the windows. The truck's lights were visible slightly to the rear, out the right side, and he heard footsteps running toward the plane. Within moments, the door flew open.

"Grant! Are you okay!?"

"I think so."

"Good. That was some landing. What did you land on the fence post for? Couldn't you see my headlights?"

"What fence post?"

"It's right there, bro', sticking out of your left wing. Looks like it practically tore the wing off."

Grant struggled out of the little airplane and walked around it in astonishment. The left wing was indeed wrapped like tin-foil around a sturdy fence post. The nose and propeller were buried against a low embankment, and the prop was bent backward at each tip. Dirt and weeds were on everything. Skid-marks left the runway 40 ft behind, and the tires gouged deep ruts into the dirt leading up to the airplane. In all the excitement leading up to the trip, Grant never figured on a crash—he'd have been better off driving.

"Better get me to a phone," he said.

BACKGROUND INFORMATION

The insurance company paid the FBO to repair the plane, then got a judge to make Grant cough up the full amount. He spent the next few years paying off a bill totaling upwards of $13,000. It was an expensive lesson in night landings. Although it is legal to land in darkness at an unlit runway, any mishap could bring judgment against you. Grant's major mistake was choosing to land into the headlights. Even on high-beam, most car lights reach little more than a few hundred feet, requiring Grant to land somewhere beyond that in near-total darkness. The wiser approach, although still risking a collision with the truck, is to land the other way, coming over the truck and attempting to touch down in the lighted area provided by its headlights. Rolling out, the airplanes' landing light would continue to light the runway ahead. Pilots flying the night mail years ago would land with little more than the light of a couple of burning smudge pots, in a similar manner.

By landing *into* the light, Grant spoiled his night vision and canceled the effectiveness of his own landing-light as he neared the truck. Looking into bright lights, his depth perception failed, causing him to think he was much closer to the truck than he really was, which, in turn, caused a panicky landing and eventual loss of control.

A better method would have been to ask the brother to drive up and down the runway, allowing the pilot to better visualize its position and length and ensuring that it was clear of obstacles and/or animals. After discussing the wind, the truck should have been positioned at the approach end of the runway, accordingly. The airport vicinity should receive careful scrutiny for potential hazards during the approach, and the pilot should plan to land over the vehicle, aiming for a touchdown point somewhere ahead of the truck, one hopes within useful range of its lights (Fig. 7-1). This way, the pilot can see the runway and its orientation while still on approach, and eventually pick up the runway with the plane's own landing light soon after passing the truck—with luck never losing sight of the runway. After the plane passes overhead, the driver in the truck might attempt to chase the airplane down the runway to provide further illumination, but this presents a certain risk of having the truck eventually collide with the rear of the airplane.

It is far easier when lights are set up on the ground properly. The following sections will follow the sequence of lights that a pilot would see when approaching an airport that is properly lighted for night operations.

Airport Beacon

From the *Airman's Information Manual,* section 2-20:

   a. An aeronautical light beacon is a visual NAVAID displaying flashes of white and/or colored light to indicate the location of an airport, a heliport, a landmark, a certain point of a Federal airway in mountainous terrain, or an obstruction. The light used may be a rotating beacon or one or more flashing lights. The flashing lights may be supplemented by steady burning lights of lesser intensity.

   b. The color or color combination displayed by a particular beacon and/or its auxiliary lights tell whether the beacon is indicating a landing place, landmark, point of the Federal airways, or an obstruction. Coded flashes of the auxiliary lights, if employed, further identify the beacon site.

Years ago, pilots followed road maps and picked a suitable field near town. If the field were farmed and they damaged crops in the landing, the farmers could be consoled with an airplane ride or perhaps a little money. As cities developed dedicated landing

Fig. 7-1. *The control towers at many controlled airports close down for the night, leaving control of lighting up to the pilot.*

fields, the fliers used beacons to navigate visually from place to place. The beacons were used as much for navigation as for marking the location of an airport. Many beacons continuously flashed a letter in Morse code, identifying the area. The beacons were scattered over the country much like VORs are today, making up a system of lighted airways, each flashing a letter identifier in Morse code. Early mail pilots would use these letters to make up acronyms for a given route, enabling them to remember the course as they flew. On a clear night, the beacons of many cities could be seen from the plane at the same time. To simplify the pilot's navigation task, some of the beacons were made to be *directional,* so as to be seen only when the pilot is roughly on the airway. Few of these directional beacons survive today.

Rotating, or omnidirectional, beacons pan their light in all directions and thus are most effective when viewed from 1 to 10 degrees above the horizon, although they may be visible from points well above or below that spread. All varieties produce some kind of flashing signal, although the method used may differ. Most common, perhaps, is the simple rotating beacon, in which the light rotates on a pedestal, giving the appearance of a flashing beacon when viewed from a distance. Another variety of beacon is a light that shines in all directions when it flashes, like the clock on your VCR when you first plug it in. Although using different methods, both beacons accomplish the same purpose, which is to provide aviators with a visual landmark of an airport or airway point. Most beacons typically flash at a rate of 12 to 30 times per minute, and about twice that for a heliport. The flashing Morse code has been superseded with a simpler, generalized color code, which follows:

- Lighted land airport—alternating flashes of white and green, sometimes just green alone
- Lighted water airport—white and yellow, sometimes just yellow
- Lighted heliport—green, yellow, and white
- Military airport—alternating two quick white flashes and green

At airports where controlled airspace is a factor when the weather is down, the beacon will often be in operation in daylight, indicating that visibility is less than 3 mi and/or ceiling is less than 1000 ft. If you see a beacon operating during daylight hours, ATC coordination will probably be a factor for using that airport. In spite of this generalization, the *Airman's Information Manual* specifies that pilots should not rely solely on the airport beacon as an indication of weather.

In the example at the beginning of this chapter, if Grant had been navigating to Manila solely by reference to beacons, he could have applied a simple rule: The red ones are dangerous, indicating a hazard (ironically, the beacon on an airplane is red), green ones indicate a land airport, and yellow indicates a water port (a difficult find in the Utah high desert). This color code has been accepted for years but has a few additions; for example, the FAA deems it acceptable to use high-intensity white flashing lights to mark an obstruction in lieu of red obstruction lights. I suppose the dazzling white lights are attention-getting enough.

After locating the airport by its beacon, and learning by reference to the color that it indicates a lighted, land airport, Grant might have adjusted his course to cross midfield and see...nothing—until he manages to turn on the *runway lights*.

Control of Lighting Systems

From the *Airman's Information Manual* (2-6):

a. Operation of approach light systems and runway lighting is controlled by the control tower (ATC). At some locations the FSS may control the lights where there is no control tower in operation.

b. Pilots may request that lights be turned on or off. Runway edge lights, in-pavement lights and approach lights also have intensity controls which may be varied to meet the pilot's request. Sequenced flashing lights (SFL) may be turned on and off. Some sequenced flashing light systems also have intensity control.

Pilot Controlled Lighting

From *Airman's Information Manual* (2-7):

a. Radio control of lighting is available at selected airports to provide airborne control of lights by keying the aircraft's microphone. Control of lighting systems is often available at locations without specified hours for lighting and where there is no control tower or FSS or when the tower or FSS is closed (locations with a part-time tower or FSS) or specified hours. All lighting systems which are radio controlled at an airport, whether on a single runway or multiple runways, operate on the same radio frequency.

Pilot-controlled lighting requires a radio or prior ground coordination. With the exception of the airport beacon, the airport lights could be off normally, including the lights aimed at the windsock. When a pilot enters an airport vicinity and keys the mike a few times, the lights come on but are preset with a timer to shut off again after about 15 minutes.

From *Airman's Information Manual* (2-7):

d. Suggested use is to always initially key the mike 7 times; this assures that all controlled lights are turned on to the maximum available intensity. If desired, adjustment can then be made, where the capability is provided, to a lower intensity (or the REIL turned off) by keying 5 and/or 3 times. Due to the close proximity of airports using the same frequency, radio controlled lighting receivers may be set at a low sensitivity requiring the aircraft to be relatively close to activate the system. Consequently, even when lights are on, always key the mike as directed when over-flying an airport of intended landing or just prior to entering the final segment of an approach. This will assure the aircraft is close enough to activate the system and a full 15 minutes lighting duration is available. Approved lighting systems may be activated by keying the mike (within 5 seconds) as indicated below:

Runway Light Activation

Key mike	Function
7 times within 5 seconds	Highest intensity available
5 times within 5 seconds	Medium or lower intensity (REIL low or off)
3 times within 5 seconds	Lowest intensity available (lower REIL or off)

Without a good chart or airport facility directory (AFD) the pilot approaching an unfamiliar airport is up the proverbial creek with regard to proper frequencies for light activation. For this reason, night flying equipment should include adequate charts and information, a flashlight, and a radio.

Lighted Wind Indicator

An uncontrolled airport may include one or more visual indicators for landing and traffic pattern direction. These include segmented circles, wind socks, wind cones, wind tees, tetrahedrons, landing direction indicators, and traffic pattern indicators. For the pilot approaching an airport in darkness, many of these visual aids will be irrelevant if they are unlighted. The pilot needs to ascertain the surface winds and choose an appropriate runway. One hopes that one of these indicators will be lighted as part of the runway environment.

In case of the windsock, many students of mine have had difficulty seeing the sock at night, even when it is lit. Since most socks are lit from above i.e., the light shines down on the sock from above, the sock will cast a shadow on the ground that is usually easier to see.

Traffic Control Light Signals

When a control tower is in operation, the tower controller operates the lights, and you won't have to turn them on. However, you are required to establish two-way communication with the controller before entering the airspace or moving about on the ground. If you cannot, because your radio is broken or, perhaps, you don't want to turn it on, the remedy involves lights again, whether day or night. The procedure involves entering the traffic pattern for the runway you intend to use and watching the tower. The tower controller will become concerned with your intentions and signal you with what is basically a high-powered flashlight.

From the *Airman's Information Manual* (4-62):

    b. Although the traffic signal light offers the advantage that some control may be exercised over non-radio equipped aircraft, pilots should be cognizant of the disadvantages which are:

       1. The pilot may not be looking at the control tower at the time a signal is directed toward him.

       2. The directions transmitted by a light signal are very limited since only approval or disapproval of a pilot's anticipated actions may be transmitted. No supplement or explanatory information may be transmitted except by the use of the "General Warning Signal" which advises the pilot to be on the alert.

    c. Between sunset and sunrise, a pilot wishing to attract the attention of the control tower should turn on a landing light and taxi the aircraft into position, clear of the active runway, so that light is visible to the tower. The landing light should remain on until appropriate signals are received from the tower.

The signal light has capability for green, red and white light, shining steady or flashing. The signals may be translated as follows:

ATCT Light Gun Signals

Color and type of signal	Movement of vehicles, equipment, and personnel	Aircraft on the ground	Aircraft in flight
Steady green	Cleared to cross, proceed, or go	Cleared for takeoff	Cleared to land
Flashing green	Not applicable	Cleared for taxi	Return for landing (to be followed by steady green at the proper time)

ATCT Light Gun Signals (Continued)

Color and type of signal	Movement of vehicles, equipment, and personnel	Aircraft on the ground	Aircraft in flight
Steady red	Stop	Stop	Give way to other aircraft and continue circling
Flashing red	Clear the taxiway/runway	Taxi clear of the runway in use	Airport unsafe, do not land
Flashing white	Return to starting point on airport	Return to starting point on airport	Not applicable
Alternating red and green	Exercise extreme caution	Exercise extreme caution	Exercise extreme caution

Once having been cleared to land, the pilot swings into a final approach and looks for the following.

Visual Glide Slope Indicators

Visual approach slope indicators (VASI) provide visual descent information to the pilot. Adhering to the on-glide-slope indications of such systems guarantees obstacle clearance during the approach in the area 10 degrees to either side of the extended runway center-line to a range of 4 nautical mi. The following explanation's diagrams are from the *Airman's Information Manual* (2-2).

The black bars depicted in Fig. 7-2, in the absence of color printing, are supposed to be red. The pilot attempts to fly the approach such that the upper lights are red, the lower ones, white. This gives rise to the statement "red over white, you're all right; red over red, you're dead."

The secret of the VASI mechanism lies in its housing (see Fig. 7-3). The box shields the lights such that you can usually see only one light at a time. There is a very narrow positional band in which both colors might be visible, giving the appearance of a pink light, but it is small enough that pilots will have difficulty actually remaining in that position on the approach.

For airplanes with elevated cockpits (think of a Boeing 747), the correct scene from the cockpit while flying a 2-bar VASI might place the rest of the airplane well below a safe glide path. With this in mind, airports which are used by such airplanes will make use of a three-bar VASI (Fig. 7-4).

This offers two potential approaches. For normal airplanes, the two upper lights should be red, the bottom white. For airplanes with high cockpits, the pilots should fly the approach such that the top light is red and the bottom two are white. For a small airplane using the upper glide-path, the pilot should expect to meet the runway well beyond the normal touchdown zone.

PAPI lights (Fig. 7-5) are a variation of the three-bar VASI in that they offer a couple of glide-slope choices; however, for normal glide-slope indications, more position information is available to the pilot, allowing for smaller corrections and a more precise approach, hence the name, Precision Approach Path Indicator.

2-Bar VASI

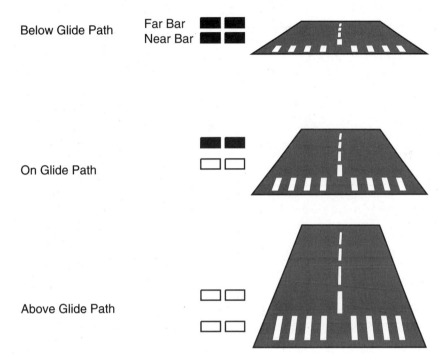

Below Glide Path — Far Bar / Near Bar

On Glide Path

Above Glide Path

Fig. 7-2. *Visual approach slope indicator.*

Side view of a typical VASI system

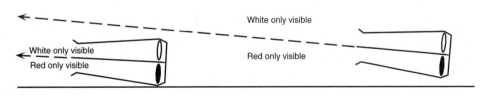

White only visible

White only visible
Red only visible

Red only visible

Fig. 7-3. *The VASI lights are housed in a box with a divider between red and white. Usually only one color is visible at a time. The two units are mounted such that a pilot on the proper glide path will see the red lights from one box and white from the other. Simple.*

3-Bar VASI

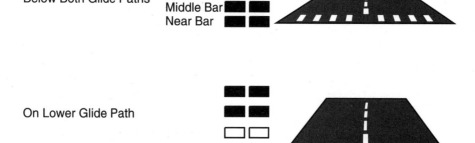

Below Both Glide Paths   Far Bar
Middle Bar
Near Bar

On Lower Glide Path

°**Fig. 7-4.** *VASI system that offers indications for two glide paths.*

PAPI

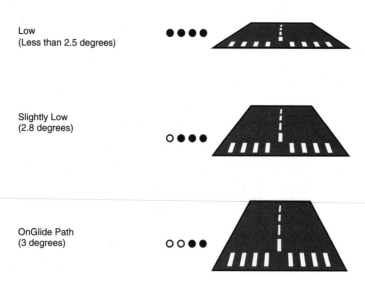

Low
(Less than 2.5 degrees)

Slightly Low
(2.8 degrees)

OnGlide Path
(3 degrees)

Fig. 7-5. *Precision approach path indicator.*

A tricolor visual approach system uses one box that contains three lights inside (Fig. 7-6). Depending on the plane's position, the pilot will see amber, green, or red light. Consistent with the other systems, seeing only red light is a no-no. Amber light indicates a high approach, and green is on glide-path.

Yet another variation on the visual approach slope indicators is the pulsating light (Fig. 7-7). In this case, it flashes white if the plane is too high, steady white when on glide path, steady red when slightly low, and flashing red when you're about to crash. There is some concern expressed in the AIM that pilots might mistake some of these flashing indications for other aircraft or obstacles on the ground, but you can do only so much with a single light.

Finally, glide-slope information might be depicted mechanically by elements on the ground, which can be as simple as painted boards on stilts, positioned so that they form a straight line when viewed from a correct position on the glide-path (see Fig. 7-8).

Once having used the glide path indications provided by one of these systems, the pilot approaching a dark runway might next see some form of approach lights, before actually seeing the runway environment.

Approach Lights

There are several different versions of approach lighting. Every version's primary purpose is to aid pilots while transitioning from instruments to visual guidance just before landing. The information approach lighting provides can be helpful to a VFR night pilot.

Tri-Color VASI

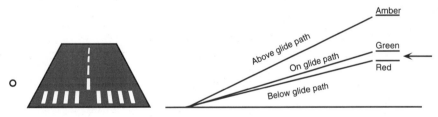

Caution: When the aircraft descends from green to red, the pilot may see a dark amber color during the transition from green to red.

Fig. 7-6. *Tricolor visual approach slope indicator.*

Pulsating VASI

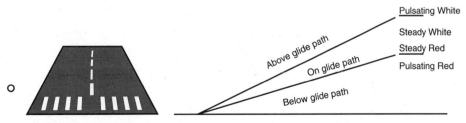

Fig. 7-7. *Pulsating visual approach slope indicator.*

Alignment of Elements

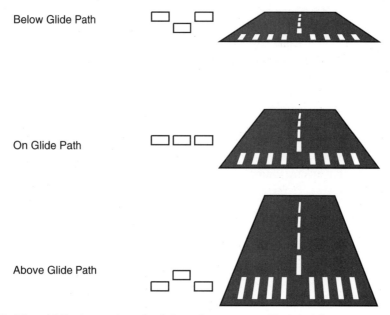

Below Glide Path

On Glide Path

Above Glide Path

Fig. 7-8. *The middle element is set back from the outer two. Their heights are set such that they appear to line up when viewed from the proper glide path.*

From *Airman's Information Manual* (2-3) on runway end identifier lights (REIL) (Fig. 7-9):

> REILS are installed at many airfields to provide rapid and positive identification of the approach end of a particular runway. The system consists of a pair of synchronized flashing lights located laterally on each side of the runway threshold. REILs may be either omnidirectional or unidirectional facing the approach area. They are effective for:
>
> 1. Identification of a runway surrounded by a preponderance of other lighting.
> 2. Identification of a runway which lacks contrast with surrounding terrain.
> 3. Identification of a runway during reduced visibility.

These things can be very bright and annoying to a pilot on short final. Most are hooked up to dimming mechanisms so that their intensity may be reduced or even shut off with a few keys of your microphone. Once you have spotted the runway lights, it would be wise to get rid of the REILs if you can.

From the *Airman's Information Manual* (2-1) on approach lighting systems (ALS):

> b. Approach light systems are a configuration of signal lights starting at the landing threshold and extending into the approach area a distance of 2400-

3000 feet for precision instrument runways and 1400-1500 feet for non-precision instrument runways. Some systems include sequenced flashing lights which appear to the pilot as a ball of light traveling towards the runway at high speed (twice a second).

At Butte, Montana, the flashing lights follow the landing pattern clear around to the downwind, to aid pilots in a circling approach for landing to the north. Butte is surrounded by mountains which pose a threat to faster planes while in the pattern. Since a wide pattern is impossible, and the mountains themselves are not lit, the approach lights attempt to provide visual guidance from the point where the airplane leaves the instrument procedure and patterns to use the runway.

Approach lights commonly look like those shown in Fig. 7-10. Notice that the spacing between the large horizontal elements is identical with all systems. Seen from the cockpit, the spacing of these bars offers some glide-path indications to a pilot who is used to looking at them, as shown in Fig. 7-11.

Runway Lights

Having finally found the runway, a night pilot may be faced with a spectacular menagerie of lights, flashing and steady, of several different colors. A lighted airport can be a beautiful, dazzling sight (Fig. 7-12). In spite of the lights, however, some practice is still needed to comfortably use them. For an illustration of the nature and design of runway lighting, we'll look at the *Airman's Information Manual* again, beginning with section 2-4:

a. Runway edge lights are used to outline the edges of runways during periods of darkness or restricted visibility conditions. These light systems are classified according to the intensity or brightness they are capable of producing: They are the High Intensity Runway Lights (HIRL), Medium Intensity Runway Lights (MIRL), and the Low Intensity Runway Lights (LIRL). The HIRL and MIRL systems have variable intensity controls, whereas the LIRLs normally have one intensity setting.

b. The runway edge lights are white, except on instrument runways amber replaces white on the last 2,000 feet or half the runway length, whichever is less, to form a caution zone for landings.

Runway End Identifier Lights (REILS)

Fig. 7-9. *These flashing lights can be very bright—and very distracting. Once the runway is sighted, most pilots turn the REILs off.*

Approach Lighting Systems

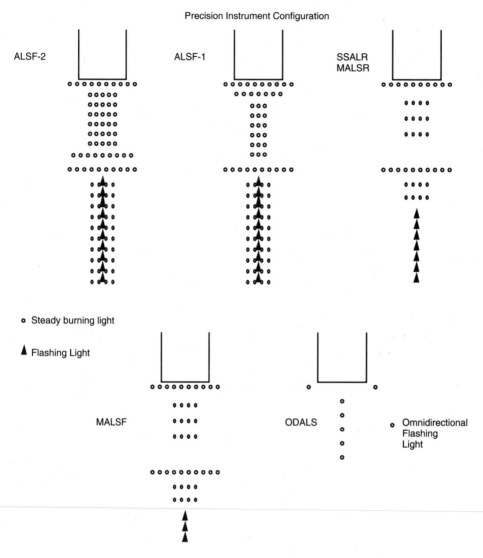

Fig. 7-10. *The actual system installed depends upon the budget and weather capabilities of its respective airport.*

    c. The lights marking the ends of the runway emit red light toward the runway to indicate the end of runway to a departing aircraft and emit green outward from the runway end to indicate the threshold to landing aircraft.

ALS system viewed from approach

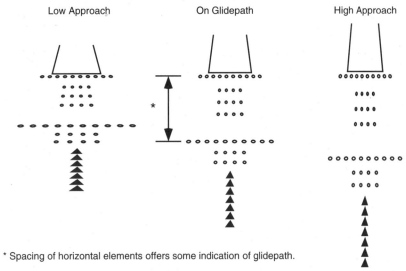

* Spacing of horizontal elements offers some indication of glidepath.

Fig. 7-11. *Approach lighting system viewed from the approach.*

Fig. 7-12. *Runway lights.*

2-5 In-Runway Lighting

Touchdown zone lights and runway centerline lights are installed on some precision approach runways to facilitate landing under adverse visibility conditions. Taxiway turnoff lights may be added to expedite movement of aircraft from the runway.

a. Touchdown Zone Lighting (TDZL)—two rows of transverse light bars disposed symmetrically about the runway centerline in the runway touchdown zone. The system starts 100 feet from the landing threshold and extends to 3000 feet from the threshold or the midpoint of the runway, whichever is the lesser.

b. Runway Centerline Lighting System (RCLS)—flush centerline lights spaced at 50-foot intervals beginning 75 feet from the landing threshold and extending to within 75 feet of opposite end. Viewed from the landing threshold, the runway centerline lights are white until the last 3,000 feet of the runway. The white lights begin to alternate with the red for the next 2,000 feet, and for the last 1,000 feet of the runway, all lights are red.

c. Taxiway Turnoff Lights—flush lights spaced at 50 foot intervals defining the curved path of aircraft travel from the runway centerline to a point on the taxiway. These lights are steady burning and emit green light.

Amid all these wonderful lights is the possibility of a displaced threshold or, in the case of the more fantastic approach lighting systems, threshold lighting is needed to identify the point at which the runway actually begins. These lights typically are steady green and run across the runway at the landing threshold or, in the case of a displaced threshold where flush lighting is not available, they run outward from the runway sides where the displaced threshold is located.

Lead-in Lights for Taxiways

I don't see the point of these lead-in lights for taxiways, when yellow taxi lines do as well, but these lights follow the taxi lines off the runway to assist pilots in locating and turning onto a taxiway after landing. The lights are embedded in the pavement at regular intervals along the taxiway centerline and are placed in alternating colors of amber and green.

Airline pilots have been known to confuse these with runway edge lights when simply taxiing along a parallel taxiway. I prefer these lights to be off, unless snow covers the taxiway (Fig. 7-13).

Need Charts

With all the lights on, the pilot faces a relatively easy task in flying a pattern for landing. The interesting part comes when a clearance is issued—"cleared to land on runway XX." If the pilot has not consulted a chart, perhaps even an airport diagram, "runway XX" might be a problem. Which one is it? The runway lighting does not light up the runway numbers, and by the time the numbers are illuminated by the airplane's landing light, the tower could be calling a violation. At large airports with multiple runways (Fig. 7-14),

Fig. 7-13. *Although sometimes confusing, taxiway lights are useful when snow covers the ground.*

Fig. 7-13. (*Continued*)

the runway lights are designed to be directional, i.e., pilots can see the runway lights well only when they are approximately lined up on final. Airplanes on downwind sometimes have a difficult time seeing the runway environment well, with lights of this type.

I cris-crossed Reno, Nevada, several times one evening unsuccessfully looking for the airport. The wealth of buildings and bright lights down there, coupled with my unfamiliarity with the area, made the airport almost undetectable in the lights below. I was talking to Reno approach at the time, explaining that I was unfamiliar with the airport and asking for a vector to help find it. They complied and I flew over the city. After a while, approach asked if I saw it yet. I said no. They said I had flown over it and issued another vector that almost reversed my course. After a time, I had flown over the airport again, and yet another vector was issued. I looked for almost 20 minutes before finding the runway, and doing so would have been much easier with a detailed chart of the Reno area or perhaps some instrument approach plates (Fig. 7-15).

TECHNIQUE

It is generally believed that the complex lighting systems that come with category 3 approach systems make the night landing easier. This is probably true. Items such as runway centerline lighting, touchdown zone lighting, and those fabulous approach lighting systems offer a great deal of perspective information to an approaching aviator. That's good, because airline pilots, who commonly make use of these systems, need all the help

Fig. 7-14. *A daylight approach into an airport with multiple runways.*

Fig. 7-15. *There are two airports visible in this photo. Can you locate them?*

they can get. With this in mind, and for the sake of simplicity, this section will focus primarily on the simplest of lighting systems, outlining a few techniques for dealing with the inherent problems of a night landing.

As in daylight, every night approach offers a different challenge. Landing at a busy airport that is equipped with low-minimum instrument approach systems is like flying over a Christmas tree; there are so many lights on and surrounding the runway that some pilots make special effort, putting up with all the ATC hassle and waiting in line, just to land there. However, the lighted runway of a lonely airport in the boondocks looks like an island in a sea of night. Pilots using that runway may have difficulty with perspective, judging distance, glide path, and even keeping their own airplane upright without reference to instruments. There are no guarantees that the runway is free of obstacles and perhaps only the most basic of approaches. No two airports are the same.

In addition, different airplanes can present a surprising challenge in a night landing. While checking out a rented C-172 one night long ago, I frightened the instructor with some really bad landings. I bounced on the nose-wheel during one of them as the instructor yelled (too late) "flare!" and attempted to pull on the yoke. It wasn't the airplane type that was causing the problem, either. I had flown a few 172s before. It was the landing light on that particular airplane. The light was aimed more downward than I was used to. On approach, I waited until the runway was visible in the light before commencing the flare. Unfortunately, in that airplane I happily plunged into a dark runway only to be surprised by the sudden appearance of the asphalt. My landings were much better when I turned the landing light off and flared by reference to the runway edge lighting.

As you sit in any airplane, you become accustomed to thousands of visual cues—how the cowling sits in relation to the horizon, where the yellow taxi-line appears to sit when the nose wheel is rolling on it, and so on. One of the most critical cues to smooth landings is the height of your eyeballs above the pavement when the wheels are on the ground. For example, the pilot of a Boeing 747, whose cockpit sits some 60 ft high, would customarily flare the airplane much sooner than the pilot of a BD-5, whose butt is barely 6 inches from the pavement when the mains touch down. The visual cues or references each pilot uses to adjust the landing approach and flare are certainly different. Most pilots look at the ground ahead of the airplane to judge precise height information while the airplane is in the flare, but there is additional information available to the sides, as well. One of the most terrifying moments for pilots checking out in a Pitts Special, for example, is when the nose comes up to flare and completely blocks the pilot's view of the runway ahead. It is equivalent to going into the flare and shutting your eyes—a little like what happened to Grant when he was blinded by his brother's headlights. After a while, though, pilots of the Pitts and other similar airplanes learn to pick up detailed height information from what is visible to the sides, and a smooth landing is once again possible. When you fly at night, similar cues are available from the lights on the runway, even when the ground itself is invisible (Fig. 7-16).

As you taxi out onto the runway, pay attention to where the lines of runway lights appear to sit as they pass away to your periphery while you look straight ahead. Keep in mind that the runway lights are often significantly higher than the runway surface and present a different picture than the runway edge lines would in daylight. You'll notice that the plane appears to "sit" lower to the ground at night, because of this. It seemed to me, as I taught night landings in a Cessna 152, for example, that the runway lights appeared to pass by at about shoulder height when the wheels were rolling on the ground. With the landing light off, I found that the students could make consistently perfect soft-field landings using only the runway edge lights as a reference. They would flare gently, easing down into the darkness of the runway while looking straight ahead, but paying attention to the view in their peripheral vision, until the edge lights appeared to pass by at the proper height and the wheels rolled on the runway. They were invariably surprised by the smoothness of their landings. After practicing with the landing light off, the additional scenery visible with the light on made their landings that much easier.

Visual Illusions

Your eyes take into account a tremendous amount of information, seeing details which you might not even be conscious of. These details are what produce your perceptions of size and distance. When a pilot approaches a runway at night, many of those little details are lacking, which may cause the pilot to make erroneous judgments of altitude, distance, and even direction.

In Fig. 7-17 your eye can be teased into thinking that one ball is bigger than the other, because of the way the fence appears. This picture fools with your perspective. For a real-world application, refer to the pictures of the runways in Fig. 7-18. You have an

Fig. 7-16. *Runway lights may offer more accurate height information than what is visible in the airplane's landing light.*

easy time discerning which runway is 150 ft wide, and which is only 75 ft wide. Take away the details on the ground, however, and you are faced with the picture in Fig. 7-19.

Which one is which? You are viewing two different-sized runways, one from twice the distance as the other. The lights might be of little help, because their brightness could be changed, or weather and environmental factors could make any of them appear closer or further away.

Gauging distance in this example could be difficult, but as your airplane approaches closer to each runway, more detailed visual cues will eventually appear—although you might be already in the landing flare by the time you see them.

Another illusion of runway width affects the pilots perception of glide-path. In Fig. 7-20, you see two runways of different widths from the same distance out on approach. The narrow runway appears longer and may delude the pilot into thinking she or he is too high on approach.

Pilots commonly perceive glide-path positioning by referencing the apparent angle of convergence of the runway edge lights. If the airplane is on proper glide path, the runway outline will appear to grow in view as shown in Fig. 7-21.

Notice that the outline of the runway appears to elongate as the airplane approaches. This is caused by a foreshortening illusion. When you see a runway from far off, it may appear shorter than when viewed from close range. A pilot who is not aware of this visual distortion may fly the airplane below a proper glide path as the approach progresses.

A pilot aware of this illusion may easily pick out the true nature of the airplane's approach path simply by virtue of experience. Another help in gauging the approach is to make sure it is a steady one, i.e., the flight parameters are consistent throughout the approach. If the pilot maintains a constant speed, descent rate, and power setting while

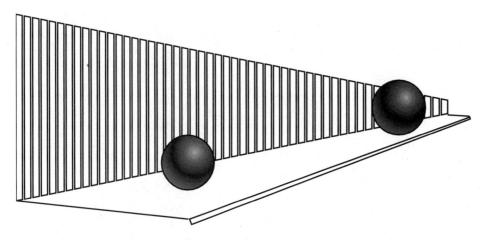

Fig. 7-17. *Perspective influences your perceptions of size.*

Fig. 7-18. *Details around an airport give relative indications of the size and distance of the runway.*

aligned with the runway, the airplane can be expected to fly a straight glide path. This can be very helpful when the true nature of the runway becomes illusory.

Things get difficult, however, when the approach itself is unstable—the pilot can't decide if the plane is high or low, and makes adjustments for both (see Fig. 7-22). In this case, the runway outline offers little help in gauging the glide path, and the aircraft instruments offer no stable reference because of the erratic and changing flightpath. So be sure to stabilize the approach if the visual cues appear misleading.

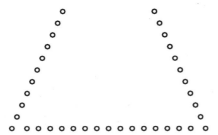

Fig. 7-19. *Without perspective information, depth perception is very difficult.*

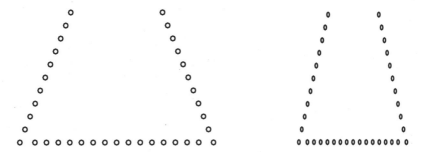

Fig. 7-20. *Slender runways may offer the illusion of a high approach.*

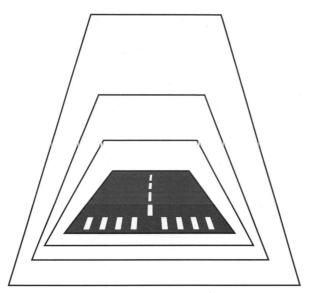

Fig. 7-21. *When viewed from a stable approach, a runway will appear to lengthen as the plane comes nearer.*

Up- and Down-sloped Runways

A pilot is not often totally devoid of visual references during a night approach. Nevertheless, even good external cues might easily be skewed by something as simple as a sloped runway. If the runway sits on a slope, for example, the picture from the cockpit when the airplane is on the correct approach might look something like Fig. 7-23. A pilot who is unaware of the vagaries of these runways could be fooled into thinking the plane is improperly positioned on approach. Correcting the picture on the up-sloped runway, for example—approaching so that the pilot's view of the runway appeared "normal"—could place the airplane perilously low, especially if there is high terrain along the approach path. The reverse is true for the down-sloped landing. The pilot's view becomes skewed and the airplane could be making a screaming dive to the landing.

With all this going on, the solution, again, is fairly simple. The pilot must pay attention to the indications inside the cockpit as well as those outside. Everyone monitors airspeed on approach, for example, but the night pilot must also monitor the power setting, rate of descent—even an electronic glide slope, if it's available. If the view outside looks normal, but the power setting is relatively high and the descent rate is unusually low for a landing approach, the clever pilot would become suspicious of a runway slope or winds.

Approaching over Obstacles

Approaching over obstacles can be frightening in the dark, because most obstacles (think mountains) are invisible. Approach paths are often altered when obstacles like this exist, meaning that VASI and other glide-slope guidance may be set for a bit higher than the standard 3.0 degrees (Fig. 7-24). If a high approach path like that is combined with, say,

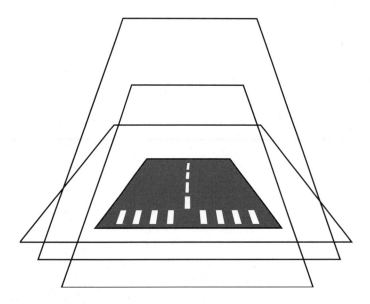

Fig. 7-22. *An unstable approach.*

Approaching a sloped runway

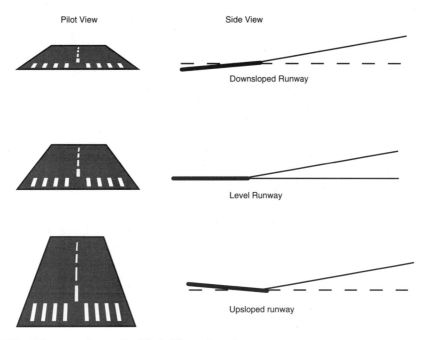

Fig. 7-23. *Pilots may be easily deluded by a sloped runway.*

a runway set on an up-slope, the night pilot on approach might be fooled into thinking the VASI is wrong or that there will be a need to go around. The illusion can be powerful.

Worse yet, the pilot could attempt to make the scene "look normal" and approach too low.

Foreground Occlusion

Foreground occlusion is a big one to watch for in mountainous regions. As the airplane descends into the runway, the pilot should pay careful attention to the lights below for any occlusion. While approaching the runway at Reno on the same flight I described earlier, the city lights appeared to wink out from below, in a peculiar way, as though a second horizon was rising beneath the plane. Reno has a little mountain just south of its main runways. The approach path for the easternmost runway passes very close to one side of it. An airplane that tracks the VASI indications to that runway may find the mountain uncomfortably close, on the right side. Since the mountain has an obstruction light on top, and the airplanes pass low and to the side, the red flashing obstruction lights are rarely noticed from close range. As we approached the airport that evening, making a right turn to line up with the runway, the mountain began to block our view of the approach lights ahead, giving the appearance of the lights going out—the mountain itself remained unseen, like a dark shadow creeping over the city. We recognized it immediately, however—the mountain is well marked on the charts—and adjusted the flight path to land without further incident.

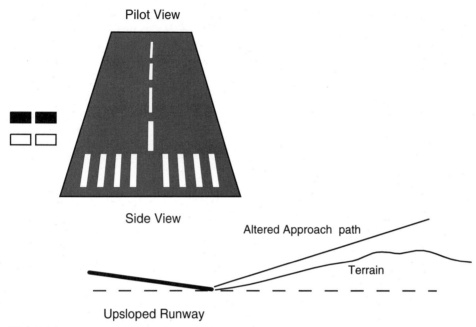

Fig. 7-24. *Sometimes approaches need to be steeper than normal.*

If a pilot notices that the lights seem to disappear in the visual foreground, an unseen obstacle is a high probability (see Fig. 7-25). Evasive action, such as an immediate climb or course change, might be required.

When descending into a terminal area, especially over unfamiliar territory, the pilot must carefully observe the lights below for any indications of the unseen. Mountains, in darkness, can look suspiciously like lakes, for example, their true nature revealed only after careful scrutiny of the way the lights change at the edges of the darkened areas they fill.

Thank Goodness for VASI Lights

As described earlier, following the recommended glide path described by visual approach slope indicator systems can do much for a pilot in preventing collisions with unseen obstacles. You remember that a VASI provides safe obstruction clearance out to 4 nautical mi along the approach path, and 10 degrees to either side. In light of the several illusions that can exist at night, simply adhering to the proper visual glide-slope indications offers comfort indeed—especially when the ground below is invisible.

Landing Performance

Nighttime conditions can greatly degrade landing performance. Consider this: You're flying the approach to a lighted runway, closely following VASI indications. VASI systems are designed to guide the pilot down to the touchdown zone portion of the runway—already 1000 ft down field. Following the VASI assures obstacle clearance but will not guide your tires to contact the runway particularly close to the threshold. Pilots may compensate for this by leaving the glide path suggested by the VASI when the runway environment is visible in the landing light, but this tends to destabilize the approach and generally results in a faster landing and/or long float prior to touchdown.

Bush pilots who make a habit of very short landings in rustic fields are usually grounded at night. Most of those rustic fields don't have lights of any kind, and the wise fliers learn to avoid them after dark. Even the most seasoned aviator will experience an increase in anxiety during a night approach as opposed to the same conditions in daylight. The increased anxiety levels tend to result in a cautiously "hot" approach, which also increases landing distance requirements.

A Cautious Approach

Without the aid of VASI guidance, night approaches should be executed with a great deal of caution. After a careful scrutiny of charts and whatever is visible in the area surrounding the runway, a pilot might yet maintain what would normally be considered "excess"

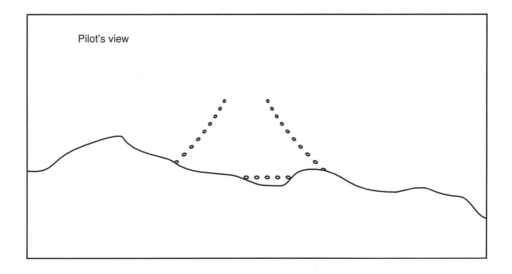

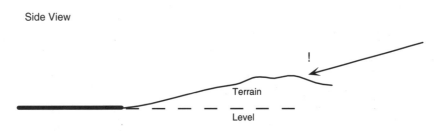

Fig. 7-25. *Foreground occlusion.*

altitude until assured of a clear approach to the runway. Many pilots choose to follow the guidelines set forth in the instrument approach charts regarding circling approaches. These charts will often state that "circling not authorized" in some quadrant of the field to ensure terrain and/or obstacle clearance in low visibility. Suggestions such as this should be of great interest to the approaching night pilot. Additional precautionary information may be obtained from the airport facility directory.

Clearing the Runway

Even though a runway might be lighted, a pilot has no guarantee that the runway is clear of obstacles, vehicles, or animals. Lighting generally makes the runway pavement even more difficult to see from the air, rather than otherwise. If cattle happen to wander onto the runway at an uncontrolled airport, pilots could easily be surprised and crash.

If the pilot cannot be assured by someone on the ground that the runway is indeed clear, i.e., "cleared to land," there is no choice but to make a precautionary pass, low enough for the pavement to appear in the plane's landing lights, and to inspect the runway personally. Perhaps the pilot might be able to observe the landing of another aircraft and thereby be assured that the runway is unobstructed—the "you go first" technique. Either way, to simply turn on the lights and land at an uncontrolled, rural airport without first ascertaining whether the runway is clear is to court disaster.

SKILLS TO PRACTICE

One of the best ways to prevent night landing mishaps is to simply pay attention—to see and be aware of everything around the airplane all the time, including what the airplane itself is doing. At night, knowing the normal operating parameters of your airplane during approach can be very helpful in detecting many of the more upsetting illusions you may face.

Stabilized Approach

Conduct several approaches in daylight. Make every effort to be consistent in your procedures, i.e., begin the descent at the same point, follow a consistent glide path, use consistent speeds, and so on. During the approaches, take notes of the "normal" descent rates and power settings required for your airplane.

When you fly at night into an unfamiliar airport, any variations from "normal" will provide clues that something might be wrong. Needing more power than usual, for example, could be an indication of a headwind or perhaps an unusually low approach. Too little power might show the opposite: a tail wind or, perhaps, an approach that is too high. Other indications in the cockpit can also be related to more serious indications outside. The point here is for you to pay attention to your airplane when you can see what is going on, and learn what "normal" is. At night, you simply try to make things as normal as possible.

Turn Off the Landing Light

This might seem scary, but it should be practiced by every night pilot. If you are uncomfortable with this procedure, get a qualified instructor to go with you. Landing lights commonly burn out. With the light off, learn to gauge the landing flare by reference to the runway edge-lights. Learn to accurately predict the exact moment when the wheels will touch the runway.

After several landings, you should be able to pick up on the visual cues available in your peripheral vision and land smoothly. Then you may turn the landing light back on and land normally. In doing so, don't forget what you just learned—make use of *everything* you see.

Stealth Landing

In *Piloting for Maximum Performance,* I suggested that pilots practice a stealth landing, i.e., fly as close to the runway as possible, in the landing configuration, without actually touching it. The procedure involves a normal approach and flare but requires the pilot to add sufficient power in the flare to prevent the airplane from actually touching the runway.

You are bound to make a mistake and get too low, and the mistake will result in a very smooth, soft-field landing. That's fine. The suggestion here is that you attempt to duplicate the procedure at night on a lighted runway, with the aircraft landing light off. As before, the goal of this exercise is to train your eye to pick up visual cues from the runway lights which will help you gauge the exact height above the runway.

FURTHER READING

Lewis Bjork, *Piloting for Maximum Performance,* McGraw-Hill, 1996.

Charles F. Spence, ed., *Aeronautical Information Manual/Federal Aviation Regulations,* McGraw-Hill, New York, 1998.

8
Emergencies

Terry set the manifold pressure for climb and pulled the props back to 2300 rpm. With 3 degrees pitch up, the airspeed steadied at 140 knots, producing a climb of 900 ft per minute with the engines rumbling away noisily outside. His eyes darted quickly around the panel, checking attitude, heading, airspeed, ball, altitude…he was conscious of it at first, but after a few moments became accustomed to the normal rhythm of instrument scan and settled down in his seat to watch the needles and wait for the world to roll by outside, unseen. Less than a minute after takeoff, the airplane had gone into the clouds. Raindrops spattered on the windshield and burst into stars like little fireworks; wet streamers tracing across the windows as though blown by a hurricane. He glanced over at the man in the right seat, who had been excitedly looking out the window as the runway lights faded away below the plane and then disappeared completely when the airplane plunged into the overcast. Two strong beams of light ebbed and stretched out ahead of the plane as clouds seemed to crash like violent waves over the cabin of the twin Beechcraft (Fig. 8-1). The sensation of speed was exhilarating but distracting. Terry flipped the light switches off, and the cockpit darkened red, and seemed to quiet. With nothing to see outside, Maurice sat back in the right seat and studied the broad instrument panel.

Maurice loved to fly, he'd volunteered for the trip to Lincoln in hopes of riding a small plane at the company's expense. As he looked at the old Beechcraft, he noted that it was showing some wear—like an old car—the paint peeling in a few places. He

Chapter Eight

Fig. 8-1. *"I need an airport with no ice and good weather...."*

clambered up steps and onto the wing, peering in the door. The bench seats were red naugahyde, seating three across, and the panel looked about 5 ft wide, covered top to bottom with gauges, switches, and levers. Terry dryly explained the door and emergency exit procedures, commanded that he buckle up—and don't touch anything on the panel. He said hardly a word after that.

Maurice watched in fascination. Everything was moving at once—indicator needles, propellers—the whole plane shook with anticipation as it taxied out to the runway. He watched carefully and wanted to ask questions about everything, but Terry looked busy. *Someday,* Maurice thought, *I'll get a license of my own.* During takeoff, the lights sped by at what had to be 90 mi/hr before the plane lifted smoothly from the runway. The ground fell away from the airplane as though the earth itself suddenly detached and fell into space, leaving the machine to rise into the clouds, blaring noisily like a bunch of trombones playing a rendition of a cattle drive. Then it was headfirst into the clouds. It rained hard. *Looks just like staring up into a shower head with the water on, except it comes fast—like a fire hose—and you don't get wet,* Maurice thought. *Man, what a rush!*

When the lights went out, Maurice began to study the panel. The pilot didn't look so busy anymore. He just sat there, fingering the controls, no expression on his face. Maurice wanted to chat.

"How do you know where we're going?" he said.

Terry sighed involuntarily, he would accommodate this enthusiastic passenger, although he preferred those who would just shut up and ride along.

174

"The radio, right there, is tuned to a navigation site near Thedford, the signal it receives guides this needle here (pointing at the VOR). We follow that."

"Aren't we going to Lincoln?"

"Sure, Thedford, Custer County, Grand Island, and finally Lincoln—they all have NAV stations along the way."

"Oh. How long before we get there?"

"It's about 180 miles away, so about an hour. You want something to read?"

"No, thanks. What's the orange light, over there, for?"

"One of the generator's out."

"Serious?"

"It's broken. " Terry's voice raised a little. " Look—this plane has two engines, two generators, two vacuum pumps, two radios—even two people inside. One generator can carry the load. It's fine."

"Oh.…Nice night."

"Yeah. Pretty smooth."

Maurice didn't want to offend the guy, just make conversation. He wondered if Terry had something on his mind—*something he doesn't want me to know.*

Terry didn't care if the guy was a little nervous. He hated it when curious passengers quizzed him while he flew. The man in the right seat seemed content enough, though, and appeared to relax. Good.

Flying was plenty for him, anyway. The weather everywhere was IFR, with ceilings not more than 1000 ft along the whole course. It was raining at Grand Island, and Lincoln was forecast to remain cloudy all night. Looked like a long flight, with nothing to see. Reaching 5000 ft, Terry pulled the manifold pressure and props back to cruise, changing the tone of noise in the cabin. They passed over Thedford. Terry set the primary VOR frequency for Custer County, tracked out of Thedford, and busied himself with housekeeping chores in the cockpit.

Maurice watched quietly as Terry sorted the charts and placed them on the middle seat, then began talking on the radio. As he talked, the noise in the cabin suddenly changed—there was a loud throbbing, which quickly went away—and the plane lurched, leaning sideways. At almost the same time, Terry dropped the microphone, grabbed the controls firmly with one hand and pushed all the levers on the dash forward with the other. The volume increased on only one side of the plane. He cursed. Maurice suddenly realized what happened—*an engine's quit!*

Maurice felt a knot growing in his stomach. He wanted to ask what happened, but Terry was throwing levers and flipping switches like some kind of crazy conductor. Something was definitely wrong. It looked really serious, a broken airplane, can't see outside, and rain—*I really don't want to die,* he thought.

"What's going on?!"

"Shut up. The left engine's quit.…We'll be okay."

Terry heard the failure coming before he noticed it on the instruments. The props came out of synch, indicating that the engines were no longer running at the same speed. As the left engine wound down, and the disparity between the propeller rpm's increased, the throbbing beat went faster and quieted. It was like an eerie drum roll before a big event which hushed as the unpowered prop turned quietly in the slipstream. Terry countered the left yaw with right rudder, tested the throttle on the left side, retarded it, and feathered the propeller. The prop on Terry's side came to a stop, its blades feathered to the slipstream. He planted the rudder ball one half to the right, trimmed the plane, and backed off the right throttle a little. It had been pushed to full power moments after the failure occurred, and since it was the only engine left, it had best be well taken care of. Throughout the emergency, Terry's eyes had been focused on the primary flight gauges, maintaining altitude, heading, and airspeed. The plane slowed, then stabilized at 120 knots, holding altitude. After trimming the controls, he scanned the panel. The right engine looked good—its temperatures sat firmly in the green and all other indications appeared normal—no telling, at this point, why the left one quit. His eyes were then drawn to the orange lights on the panel—there were two.

The good generator was on the left engine. Failed engine meant failed generator. The generator on the good engine didn't work. They were down to the battery, and in the time required to cage the failed engine, the old battery was nearly exhausted and the lights began to dim. Terry picked up the microphone and tried to call center. No response. He turned down the squelch and heard a quiet hiss over the cabin speakers. The radio was already dead. He remembered looking at the DME for a distance to Custer VOR—22 miles, and made a quick calculation. He fished around in his chart bag and found a penlight. Putting the light in his mouth, he switched off all of the avionics, then the cabin lights, and finally the master switch—flying only by the glow of the pen light in his teeth.

Looking to his right, Terry could see that the passenger was scared, his eyes wide in the darkness....

Maurice watched in horror as Terry turned off all the lights, including the orange "warning" lights—darkening the cabin and panel completely. The propeller out there was stopped dead, they were in clouds, and Terry had that flash-light in his mouth, looking like some sort of red-neck chewing on a high-powered cigar, it's light reflecting back in his face, reminiscent of a goofy campfire trick. For all his sudden anxiety, Maurice began to get angry.

"What the heck did you do that for?!"

"Because we lost the generator. We don't have *any* power, except the battery. I can't leave the lights on, or *that* will wear down too; so I had to turn them off."

It sounded logical enough, but scary. They were flying through darkness, *in darkness,* and who knows where they're headed? Where *are* they headed?!

"You turned off the radios, too?"

"Yeah."

"So, how do you know where we're going now?"

"Custer county is 20 mi away. We're flying at 2 mi a minute; that's 10 minutes in this direction (pointing at the compass). After we fly that far, we'll turn a radio back on, get another fix and probably land at Grand Island."

"Is it serious?" asked Maurice, quietly.

"Yeah. It's dang serious."

Oh, man. He thought, sinking into the seat. Maurice could hardly stand the suspense. *How could a guy just sit there like that with a flashlight in his teeth, when it was obvious that we're about to die? He said it was serious, but I know it was worse than that because he kept telling me to shut up. We're probably crashing right now, and he won't tell me anything....*

Terry could feel the poor guy's discomfort—he didn't feel so good himself, but that guy was *scared.* The last thing he needed would be a panicked passenger. Terry decided to keep the guy busy, help him calm down....His thoughts quickly returned to the plane; ATC would want to know about the emergency, but the radio was already gone. By the time they get near Custer, the DME probably wouldn't work, either. If the rain was freezing at Grand Island, the plane would come down—it couldn't carry any ice with only one engine. Terry thought furiously for a solution to the problem. The land was pretty flat, so they could dead reckon all night, but finding a runway in IFR conditions would be impossible without a radio. The battery was probably gone—it probably couldn't power the NAV equipment too well, let alone crank the engine. The engine! If the engine would turn, they'd have a generator again. An idea started to come together....

A few minutes later, Terry figured they were somewhere near Custer County VOR. He flipped on the electrical systems and...nothing happened. He saw Maurice start in his seat.

"Why won't the lights come on?"

"Battery's dead. Look, I need your help. Hold this flashlight so I can see the panel— I need to talk on the radio."

"How can you use the radio with no electrical power?"

"We're not done yet. We've got an accumulator."

"A what?"

"An accumulator. It stores pressure so that we can unfeather the prop."

"To start the engine?"

"No. But it'll let the prop turn in the wind, and that will drive the generator."

Terry moved the prop lever out of feather and watched the blade angles change out the window. Suddenly the propeller disappeared into a blur and soon after, one orange light appeared on the panel. Terry had to lower the nose to maintain speed.

"We've got power for a few minutes—you can turn the flashlight off."

"Great! But why for only a few minutes?"

"Can't hold altitude with the prop turning outside. We'll get a good NAV fix, then feather the prop and go back to the flashlight."

Okay, he got the lights to come on again, but now we're going down. Maurice was at once impressed and afraid, as he toyed with the flashlight in his lap, grateful for something to do. It was nice to see again, but very uncomfortable to be in a descent. He found the altimeter on the panel and watched it closely. It did not move very fast. *We're still alive...for now,* he thought.

Terry called ATC and declared an emergency, advised them of the radio and power situation and requested an airport with no ice and good weather. ATC said that low ceilings were everywhere, but it wasn't raining at Grand Island anymore. Terry said thanks and that he'd call back in 10 minutes. He took a fix off the Custer VOR, turned everything off again, and feathered the prop. The panel went dark, except for the glow cast by the flashlight in Maurice's hand. It was 80 mi to Grand Island. The airplane had settled to 2500 ft, and Terry had to slow to 90 knots for the climb—about an hour to go. It took 25 minutes to reach 5000 ft again—a T-bone couldn't climb too well with only one engine. Maurice was relieved to see that the plane could indeed climb—although very slowly— once the prop was feathered, but the suspense was unbearable.

Terry unfeathered the prop again and got a fix on Grand Island. With the cabin lights on, he selected an ILS approach chart and contacted ATC.

"Twin Beech _____, glad to hear from you, say flight conditions, altitude and intentions."

"IFR, 4000 and descending. We'll need the VOR/DME to Grand Island, and we'll need to be a bit high, starting the approach—we can't hold altitude while we navigate."

"Roger that, cleared for the approach, direct Grand Island VOR, maintain altitude at your discretion."

"Roger. Call you back in fifteen minutes."

Terry feathered the prop and began to climb again. Fifteen minutes later—upon reaching 5000 ft—he set the NAV receivers for the VOR approach and unfeathered the prop for the third and final time. They flew over the initial fix with altitude to spare, while watching the DME and cross-checking the suggested altitudes on the approach. Terry extended the gear and flaps, trying to get to minimum descent altitude a little before the missed approach point. He figured that they would be committed to land while still 500 ft above the airport. The plane could climb on a missed approach, but would have to dead-reckon through the procedure without NAV equipment, possibly to collide with tall obstacles....

Maurice watched the descent with horrible fascination. He counted every turn of the altimeter, knowing that the plane was descending relentlessly toward *what?* His heart began to pound faster. At his point, he had nothing to do, but sit quietly and wait until Terry flew the plane to the ground, one way or another. Terry was very busy with his charts, a picture of concentration. He flipped the landing lights on to reveal the same gray violence of clouds outside. Soon after, lights appeared below. At first it was just a few lights below the plane, then Maurice could make out some details further ahead. He sat up and looked, leaning toward the window, trying to see the ground in front of the plane. It was a runway! He couldn't believe it! *That guy found a runway! We're saved!* He practically shouted.

Terry landed on the wet asphalt and began to taxi in. When the engine was shut down Maurice leaped out of the plane, danced around on the ramp and yelled "We made it!" He grabbed Terry in a big bear hug;

"You saved my life, man!"

BACKGROUND INFORMATION

Emergencies in darkness carry a greater weight than similar problems in daylight. In Terry's little adventure, for example, a simultaneous engine and electrical failure would have certainly been a problem in daylight, but the need for a flashlight made it that much more difficult, in darkness.

If the plane were not a twin, an engine failure would have caused an immediate descent and emergency landing. Breaking out of the low cloud ceiling in darkness, the pilot in that case would be faced with a monumental problem in the dark that might be greatly alleviated in daylight. An electrical failure in a single would also be severe, although the emergency landing would not be quite so immediate. A pilot could dead-reckon by compass, clock, and partial panel until the tanks run dry, hoping to find VFR conditions. It's still not a nice prospect, but at least the landing is delayed, perhaps for the better.

Engine and electrical failures aside, there are numerous things that could cause an emergency. Simply forgetting a flashlight might make it impossible for the pilot to read vital charts in darkness. In that case, there is little choice but to declare an emergency and ask for specific directions from ATC.

Landing lights commonly burn out. For a pilot unpracticed at landing without one, this simple problem could potentially be disastrous.

Unseen obstacles or animals could block the runway. A pilot who can clearly see runway lights might attempt to land, assuming the runway is clear. It would be a real surprise to find a deer, frozen scared in the airplane's landing light. In that case, there would be little the pilot could do except hit it. What if it were something larger, like a horse, cow, or moose?

Piper PA-28-181
20 June 1993
Provo, Utah

> The aircraft struck deer on the runway, during landing ground roll, [at night]. No injuries (to the occupants of the plane).
>
> Probable cause: The wild animals on the runway. A factor relating to the accident was the dark night conditions.

Although daylight does not guarantee that the runway will be clear, the pilot would then more easily notice animal obstructions in time to take corrective action (using a rifle?).

SOME NOVEL SOLUTIONS TO THE "NIGHT QUESTION"

Since most night emergencies center around the fact that it may be difficult for the pilot to see, the following creative ideas are available for your perusal.

Night Vision Goggles

I have spoken with a few army personnel who commonly make use of night vision equipment. This is rather expensive equipment that can detect and enhance the nighttime scenery so that a soldier or helicopter pilot can operate in nearly total darkness. I wonder if equipment like this might be useful to a pilot who is faced with the need to make an emergency landing in darkness.

Invasion Flares

The military occasionally uses very bright flares to shine the way in darkness when they carry out operations involving the need for many to see at once, such as a night parachute drop, invasion, or artillery barrage. The flares are typically dropped from an airplane or fired from a gun, such that they burn while floating slowly to the ground on a parachute. They can be as bright as the stadium lights for a Yankees ball game, lighting up the ground below like midday.

When the Beechcraft Bonanza was designed, lighted runways were not nearly so common as they are today. The company offered an option to buyers of the new V-tailed airplane that included the installation of a compartment containing flares for night landings. The pilot could eject a flare over an airport, temporarily lighting up the area so that a landing would be possible. The idea worked fine but was discontinued over fears that pilots, happily playing with flares, might cause other things to burn on the ground—like forest fires.

Smudge Pots

These are cheap landing lights. Someone on the ground burns contained fires around the perimeter of the runway, and the pilot references these to land. Obviously, it requires coordination from someone on the ground. In Lindbergh's day, the fires were considered costly in fuel, and the mail company urged pilots to land without them whenever possible.

Super-powerful Landing Lights

This is an alternative to invasion flares that is used commonly by nighttime crop-dusters. If the airplane can develop the required electrical current, huge landing lights could be fitted which may be adequate to light up the ground well enough from a distance so that a normal approach and landing is possible at night.

Spotlights

One of the difficulties with landing lights is that they are mounted to the airplane in a way that they cannot be aimed without turning the plane. A controllable spotlight is the solution to this problem. Similar to the lights on highway patrol cars, but much more powerful, spotlights are used by police and filming helicopters to light up the ground. They can be amazingly bright. For a pilot faced with an emergency landing, a spotlight could enable the pilot to examine the terrain below and around the airplane before actually committing to land.

The downside of powerful lights of any kind is that the electrical power required to run them usually comes from an engine, and an engine failure is usually the reason for the emergency landing. Nevertheless, bush pilots and others using unlit runways have to get their own lighting somewhere—why not bring it along?

Handheld Equipment

Since the advent of portable aviation radios and navigation equipment (GPS), the problem of lost communications and direction because of an aircraft electrical failure should be solved. Unfortunately, the day you forget to bring your flight bag is the day the alternator fails and you're caught without your equipment. Nevertheless, handheld equipment can operate independently of the airplane's electrical system and offers a great advantage when the chips are down. In Terry's case, a portable radio and navigation unit would have eliminated the need for the periodic descents, enabling the airplane to remain safely at altitude and cover ground at a higher speed. The emergency not only would have been relatively benign, but it would have been over in less time. Handheld equipment is cheap, considering the potential cost of an accident.

Remarkably, the simplest, commonly used handheld item is a flashlight—don't fly away at night without one (Fig. 8-2).

Parachutes

This idea is mentioned in Chap. 2 but bears repeating here. A parachute, like an ejection seat, is a last resort. If you leave the airplane, the plane could fall on unsuspecting people below. The plane will almost certainly be destroyed. Upon leaving, you will be suddenly without much of the survival equipment in the plane and still may face a difficult night landing (although slower), feet first. Your reasons for jumping out of the plane probably stem from the fact that you could not see the ground. From a parachute, you *still* won't see the ground, and what you don't see might hurt.

In many cases, particularly with the lighter and slower airplanes, you might be better off with the classic crash landing technique. Minimize the forward speed and sink rate, tighten your safety belts, crack the doors, and hope for the best. It's going to be scary, either way.

Redundant Equipment

Mechanical redundancy is the most widely accepted way of combatting the emergency landing—carry two or more engines, extra generators, redundant lights—even extra pilots. Airlines are designed to remain in the air with multiple emergencies, hopefully preventing an emergency landing from happening at all. However, extra engines and systems are heavy, and the planes that carry them are fast and expensive. If they ever do actually crash, the damage from the heavy, fast moving landing is usually immense. You'd have a much easier time surviving a crash into the unseen while flying an ultralight than a Boeing 747.

Nevertheless, preventing the crash in the first place is usually better than surviving one, after the fact. I personally side with the redundancy crowd.

Go–No Go Decision

As in any flight, the decision to depart or fly at night marks the starting point of a successful flight, or a crash. Your survival could hinge on judgment exercised before you

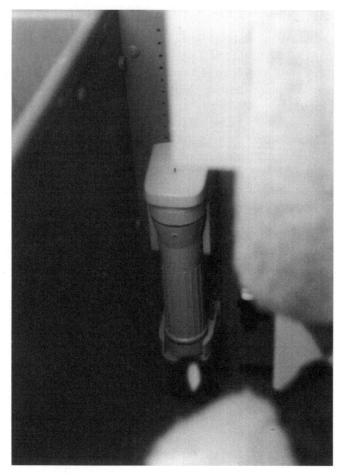

Fig. 8-2. *If you remember to bring a flashlight, make sure it's handy in the cockpit.*

even get in the plane. In Chap. 1, you were presented with several pros and cons regarding night flight. Many of those were examined with greater detail in the chapters that followed. Here you are in the final chapter, having been briefed on the challenges of night flight. As you consider a night flight, on what conditions would *you* base your personal go–no go decision?

Because pilots are easily swayed by ego or the need to be someplace, it would be wise for you to consider the night question when neither of these personal factors are involved. For Terry, flying with two engines and only one generator was not a big deal, initially. He handled the emergency situation skillfully, but perhaps, after his adventure, his personal equipment limits would be different.

As you sit there, planning to be up in the night, without ego, need, or even, perhaps, opportunity, to fly, your judgment is probably best. Take that opportunity to consider and establish some personal guidelines about flying in the dark. Ask yourself some questions—what are your personal limitations with regard to flying in the dark? Would you go if the plane's landing light were gone? Would you go without lights at all? Would you require a single or a multiengine airplane? Two alternators or generators, or one? What about the weather? Is marginal VFR acceptable for you? Would you fly an airplane without attitude instrumentation at nighttime? How much fuel reserve would you personally take? Are you current in night operations? How much practice do you feel you need to become current? Is there terrain over which you would not fly at night that you would tackle in daylight? What is the minimum weather you need to fly night VFR? At what point would you plan to go IFR? What type of ground facilities do you require? How long should the runways be? Do you need VASI-type guidance?

There are a great many more questions you could ask, and, in the end, few flights will conform exactly with any of them. In considering questions like these beforehand, however, you may clearly define your personal concerns and use that to make your preflight and in-flight decisions. Hopefully, your judgment will be adequate to prevent emergencies from happening at all.

TECHNIQUE

Since the foregoing involves things that are not readily practiced (like parachuting out of your airplane, at night), the best I can offer is that pilots think about and plan for their options in a night emergency—even after the emergency has occurred. A little planning will do a great deal for a pilot in preparing to meet a challenge in darkness. In Terry's case, he worked hard during the thick of the emergency, but he kept his head, considered his options, and came up with a solution that worked. May all pilots do the same, when their time comes.

SKILLS TO PRACTICE

An electrical failure in total darkness may leave you groping for the proper switches. This drill is used commonly in military training circles to prevent mistakes and has direct benefit here. Sit in your cockpit blindfolded and touch the switches as you would in a flight sequence. Find the master switch, magnetos, mixture, throttles, gear flaps, trim, and so on (Fig. 8-3a). Can you readily find the circuit breaker to disconnect the autopilot or electric trim (Fig. 8-3b)? Can you find the cockpit panel light controls without looking (Fig. 8-3c)? Can you reach in the glove box and quickly retrieve a flashlight? If you practice long enough, your body will become conditioned to reach for each knob correctly, without the benefit of sight, enabling you to operate better in a night environment.

FURTHER READING

Appendix A

Fig. 8-3a. *Can you reach the correct knobs in your cockpit without looking*

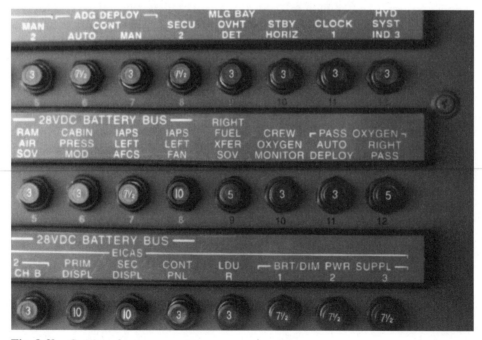

Fig. 8-3b. *Continued*

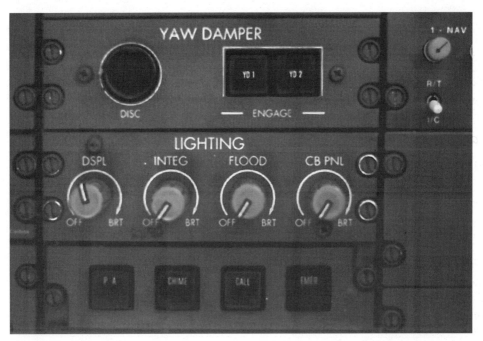

Fig. 8-3c. *Continued*

There are many ways to "go bump" in the dark

A perusal of the NTSB's impressive collection of aviation statistics revealed many night-related aviation accident synopses. Some representative selections from that database follow. I have made a few unofficial observations of my own regarding these files—I am personally acquainted with the facts and people surrounding a few of them. Although much of this material is rather sobering, it is included here to help you understand the risks involved in night flying. Because the accidents referenced here happened at night, you may determine that night flying is unusually dangerous. For the sake of unbiased judgment, however, keep in mind that the vast majority of aviation accidents happen in the daytime. Flying in darkness offers different risks than daylight flying, but it is not necessarily more dangerous in itself.

The pilots involved in night accidents run the gamut from experienced airline pilots to students. There are even a few bus drivers and ramp personnel who contributed to the statistics. You may draw your own conclusions about how "the other guy" made bad choices or was terribly unlucky, while remembering that, someday, you could wear the other guy's shoes.

Appendix A

Table A-1 A 10-Year Synopsis of Accident Statistics

				Day			
Year	No. of aircraft involved in accidents	No. of accidents	No. of fatal accidents	No. fatalities	% fatal	Average no. of fatalities per accident
1986	2307	2268	351	632	15.48	1.80
1987	2245	2210	358	732	16.20	2.04
1988	2149	2120	353	606	16.65	1.72
1989	2047	2016	362	881	17.96	2.43
1990	2060	2031	372	626	18.32	1.68
1991	1983	1946	348	674	17.88	1.94
1992	1881	1861	362	688	19.45	1.90
1993	1844	1819	315	554	17.32	1.76
1994	1810	1788	331	747	18.51	2.26
1995	1874	1847	331	584	17.92	1.76
1996	1700	1672	255	415	15.25	1.63
Totals	21900	21578	3738	7139	17.41	1.91
				Night			
1986	329	328	122	258	37.20	2.11
1987	311	306	104	221	33.99	2.13
1988	280	279	104	214	37.28	2.06
1989	275	274	97	406	35.40	4.19
1990	238	234	98	273	41.88	2.79
1991	246	242	96	208	39.67	2.17
1992	231	231	99	235	42.86	2.37
1993	237	237	94	208	39.66	2.21
1994	243	238	92	204	38.66	2.22
1995	215	214	79	144	36.92	1.82
1996	148	146	49	210	33.56	4.29
Totals	2753	2729	1034	2581	37.89	2.50

The data in Table A-1 reveal that most flying is conducted during the day. Night accidents make up just over 12 percent of the total but almost 36 percent of the fatalities. From this, we can look at the odds—for pilots up in the night, there is half the risk of colliding with another airplane and about twice the risk of a fatality. This would seem to reflect that

1. Other (lighted) airplanes are easier to see at night, and;

2. Unseen ground might occasionally be difficult to avoid.

Here are a few lessons offered through the experience of other pilots, during one recent October.

ACCIDENTS AT NIGHT, OCTOBER 1997

Denver, Colorado

On October 1, 1997, at 0436 mountain daylight time (MDT), a Boeing 727, operated by Ryan International Airlines, Inc., was involved in a ground collision with an airport shuttle bus while taxiing for takeoff. The airline transport-rated captain sustained serious injuries, the airline transport-rated first officer received minor injuries, and the airline transport-rated flight engineer was not injured. The shuttle bus driver received minor injuries. Two passengers aboard the bus were not injured. Visual meteorological conditions prevailed.

Preliminary information indicates the shuttle bus was en route from the employees' south cargo ramp. According to the first officer, [the 727] had been cleared to taxi "right side out" and to hold short of the taxiway. There is a stop sign at the intersection. The [bus] driver reportedly stopped before proceeding across the ramp. A bus passenger saw the aircraft approaching from the left and yelled three times for the driver to stop. The driver later said he thought the passenger was referring to an aircraft off to the right side. The first officer said the navigation lights, rotating beacon, and runway turnoff lights were operating. He saw "something dark" off to the right and yelled a warning to the captain. He did not see the bus headlights. The flight engineer said she saw the bus, but "the bus was hazy. It was not clearly discernible." She did not see headlights or interior lights on the bus. The captain could not be interviewed. The left and right main tires left skid marks measuring 22 and 24 ft, respectively. The nose tire was pushed to the left [leaving] a skid mark 16 ft in length. The nose section of the airplane was destroyed. The left front portion of the bus was damaged and the windshield was knocked out.

It surprises me that a relatively small vehicle, like the bus, could do so much damage to a Boeing 727—the bus driver received only minor injuries, as opposed to the jet captain, who was hospitalized. Perhaps it stems from the fact that the airplane is made of aluminum, to be lightweight, and the bus is probably steel. So, if you consider landing on a busy road, the cars and trucks there will fare better than your airplane.

Lake City, Michigan

On October 1, 1997, at 1830 eastern daylight time (EDT), a Mikowski Challenger II, owned and piloted by a private pilot, was destroyed when it collided with the terrain shortly after takeoff. The pilot reported serious injuries. Visual meteorological conditions prevailed at the time of the accident. The flight departed [the airport] at 1828 EDT.

Nebraska City, Nebraska

On October 1, 1997, at 1945 central daylight time (CDT), a Piper [Arrow], piloted by a private pilot, was substantially damaged during an off-airport forced landing following a total loss of engine power. Visual meteorological conditions prevailed at the time of the accident. The pilot reported no injuries. The flight departed Iowa City, Iowa, at 1725 CDT.

Cross City, Florida

On October 1, 1997, about 2044 EDT, a McDonnell Douglas DC-9, registered to American Airlines, Inc., experienced an in-flight encounter with a weather cell during cruise flight near Cross City, Florida. Instrument meteorological conditions prevailed at the time, and an IFR flight plan was filed for the scheduled, domestic, passenger flight. The airplane was not damaged, and the airline transport-rated captain, commercial-rated first officer, and 89 passengers were not injured. One flight attendant was seriously injured, and two flight attendants and one passenger sustained minor injuries. The flight originated about 1755 central daylight time from the Chicago-O'Hare International Airport, Chicago, Illinois. According to the flight safety department of the airline, during cruise flight at flight level 330, the flight encountered what was reported as "clear air turbulence" which resulted in 1 jolt of reported moderate turbulence. One of the flight attendants broke her wrist as a result, and the flight was continued with an uneventful landing about 24 minutes later. The seat belt sign was not illuminated at the time of the occurrence. Preliminary review of satellite images about 29 minutes before the accident indicate an isolated cell slightly to the west of the jet airway in the vicinity of the occurrence. Further review of the images about 1 minute after the accident indicates a cell directly over the airway with a larger cell south and to the right of the airway.

Sometimes nighttime weather can catch pilots by surprise—they reported "clear air turbulence," indicating that they didn't see it coming in the darkness, catching the cabin crew unprepared. Notice that this is, so far, just one day, not too long ago. The mishaps are typical for any day of the year, however—these things are not exactly rare.

Midland, Texas

On October 3, 1997, at 1916 central daylight time, a North American Harvard Mk IV, lead airplane in a formation flight of four, collided with a North American SNJ-5, number two in the formation, during landing roll at Midland International Airport, Midland, Texas. Both aircraft were substantially damaged, and both pilots were not injured. VMC conditions prevailed, and a flight plan was not filed for the local flight.

According to the FAA inspector, during landing roll on runway 34L, the lead aircraft passed taxiway Foxtrot. As the lead aircraft proceeded to make a 180-degree turn to exit the runway onto taxiway Foxtrot, the right wing of the number two aircraft struck its vertical stabilizer, separating it and the rudder from the fuselage. The number two aircraft's right wing was substantially damaged. Both aircraft taxied to parking.

Albuquerque, New Mexico

On October 4, 1997, approximately 1900 mountain daylight time, a Cessna 210L was substantially damaged during an inadvertent wheels up landing at Albuquerque, New Mexico. The private pilot, the sole occupant aboard, was not injured. VMC conditions prevailed, and no flight plan was filed for the personal flight. The flight originated at Santa Rosa, New Mexico, approximately 1800. This event was originally classified as an incident and upgraded to an accident after the airplane was inspected by a FAA airworthiness inspector. The pilot reported that while on final approach to runway 30, he made a final prelanding check. His attention was distracted by an annunciator light on the GPS (global positioning system) that had suddenly illuminated. The pilot then heard the gear warning horn. He failed to ascertain that the landing gear was down and locked and inadvertently landed wheels up.

Small, inconspicuous lights on the panel might be distracting at night, perhaps more so than in daylight. Many instrument lighting installations feature dimmer switches, to prevent this.

Orlando, Florida

On October 8, 1997, about 2132 EDT, a Beech 18, operated as a cargo flight, impacted with the ground near Orlando, Florida. VMC conditions prevailed, and an IFR flight plan was filed. The airplane received minor damage. The commercial pilot was not injured. The flight had departed Melbourne, Florida, at 2050. The pilot had a problem lowering the landing gear, and elected to land gear up on runway 18R.

Although this one happened at night, it would have had the same result in the daytime. As you will see, many of these night mishaps would not have been appreciably different in the day. Stuff like this may happen anytime.

Leadore, Idaho

On October 10, 1997, approximately 1930 mountain daylight time, a Piper Archer impacted the terrain about 8 mi east of Leadore, Idaho. The private pilot and his passenger received fatal injuries, and the aircraft was destroyed. The personal pleasure flight, which departed McCall, Idaho, about 1 hour prior to the accident and was en route to Blackfoot, Idaho, was reportedly operating in instrument meteorological conditions at the time of the accident. No flight plan had been filed, and the ELT, which was activated by the impact, was the primary means by which the wreckage was located. Witnesses reported seeing the aircraft flying under a low overcast about 10 to 15 mi west of the accident site about 15 minutes prior to the crash. According to these witnesses, the aircraft was 300 to 500 ft above the ground, and the ambient light was just turning from dusk to dark. They said that although it was not snowing where they were, there was snow falling on many of the hills around the valley through which the aircraft was flying. At first, they thought the aircraft was trying to land on the road so they pulled off to the side. But, the aircraft continued on to the east, toward Gilmore Summit, and eventually disappeared from sight.

In retrospect, flying visually into darkening, IFR conditions, landing on the road might have been a better choice. You might refer to the story that begins Chap. 6.

Knoxville, Tennessee

On October 11, 1997, about 2051 EDT, a Piper Cherokee 180, operating on a personal flight, crashed in the Tennessee River, about 1 mi west of the Downtown Island Airport.

Appendix A

VMC prevailed, and no flight plan was filed. The airplane sank and was destroyed. The private-pilot and two passengers were fatally injured. The flight was originating at the time. The occupants had arrived in Knoxville earlier in the day to attend a football game. According to the airport personnel the three occupants had arrived back at the airport about 2030, after the control tower had closed. The pilot had line service personnel put 24 gallons of fuel in the airplane, and, according to the line service manager, this "filled the tanks." The airplane was seen by several witnesses climbing in a westerly direction, about 200 ft above the ground (AGL), in a nose-high attitude, wings rocking, and lights on. Witnesses said the airplane turned, went into a nose-low attitude and impacted the water. Several witnesses said the airplane's engine was "revving" and sounded "loud," before impacting the water. One witness said the engine sounded like it was at "full power." The winds were reported to be calm at the time of the accident.

Piqua, Ohio

On October 15, 1997, at 1710 EDT, a Cessna 152 was substantially damaged during landing at the Piqua Airport (I17), Piqua, Ohio. The student pilot was not injured. VMC prevailed for the local flight. No flight plan had been filed for the solo instructional flight. According to the pilot, after returning from the practice area, he flew over I17 and observed the tetrahedron favoring the use of runway 8. The pilot entered a downwind for an approach to runway 8. He further stated: "…There were minor corrections on approach. I touched down, and then my left wing felt like it was going to flip over. The aileron correction brought the wing to touch back down, but the plane veered off the left of the runway. There was a hump that ran parallel to the runway that threw the plane back aloft. I tried to keep the nose up, but I guess my speed was too slow, sending the plane down nose first." Winds reported at an airport about 16 mi south of I17, at 1651, and 1751, were from 350 degrees at 8 knots, and from 10 degrees at 12 knots, respectively. Examination of the airplane by a FAA inspector revealed no evidence of preimpact abnormalities.

This one happened while it was yet light enough to see, although the sun was low on the horizon. Considering the wind reported from the airport nearby, is it possible that the student pilot misread the tetrahedron and landed in a crosswind? Would an error like this be easier to make in darkness?

Winston, Montana

On October 18, 1997, approximately 2130 MDT, a Bell 47G-3B, operated and flown by a private pilot, sustained substantial damage during a hard landing following a forced landing near Winston, Montana. The pilot and passenger were uninjured. VMC bright night conditions existed, and no flight plan had been filed. The flight, which was personal, was returning from Townsend, Montana, en route back to Helena. The pilot reported to an FAA inspector that while climbing through 4800 ft above the ground the helicopter yawed momentarily and then the engine abruptly ceased developing power. The pilot transitioned into an emergency autorotation and was unable to regain power during the descent. Just prior to landing the pilot maneuvered right to avoid trees and touched down hard on down-sloping terrain during which the left skid buckled and the main rotor severed the tail boom.

A successful autorotation requires precise timing on the pilot's part. Although the bright night conditions allowed the pilot to see and avoid trees, diminished depth perception probably contributed to the hard landing. Nevertheless, "any landing you can walk away from...."

College Station, Texas

On October 19, 1997, at 0216 central daylight time, a Mooney M20 airplane was substantially damaged following a loss of control while executing a go-around near College Station, Texas. The noninstrument-rated private pilot and the front seat passenger sustained minor injuries, and the two rear seat passengers were seriously injured. VMC prevailed for the night cross-country flight for which a VFR flight plan was filed. The flight departed from Oklahoma City at 1155. The pilot told the FAA inspector that he "was too high and too fast on his approach" to runway 16 at Easterwood Field Airport, so he elected to execute a go-around when the airplane was approximately 40 ft above the runway. The pilot further stated that "the airplane stalled after he applied full power and retracted the landing gear and flaps." The night cross-country flight originated at Manhattan, Kansas, where the four college students aboard the airplane attended a Kansas State-Texas A&M football game. The flight landed at Oklahoma City for fuel near midnight. Examination of the airplane by the FAA inspector confirmed that the right wing and the engine mounts sustained structural damage. The mixture control lever was found "approximately half-way out" from the full rich position.

A pretty late flight after a, no doubt, exciting football game—after all the yelling and enthusiasm that goes into a good ball game, the flight home might be rather tedious. It appears as though the mixture was not enriched during the landing approach to prepare for a possible go-around, as is commonly practiced. Also, a Mooney has very powerful trim forces (the entire empenage moves)—with the nose trimmed up sufficiently for approach at suitable power settings for landing, suddenly applying full power causes the nose to pitch up aggressively, which demands rapid and forceful attention from the pilot. If the pilot were tired and flying in dark conditions where the attitude might not be so apparent, things like this may happen. (This is simply an observation, not a statement of fact as to what happened in this particular case.)

Thompsons, Texas

On October 19, 1997, approximately 1830 CDT, a Cessna 182P airplane was substantially damaged during a forced landing following a loss of power, 4 mi west of Thompsons, Texas. The flight instructor and private pilot were uninjured as a result of the instructional flight. VMC prevailed, and no flight plan was filed for the flight, which originated from Houston-Southwest Airport, at 1745. In a written statement to the FAA inspector, the private pilot stated that he was receiving instruction on unusual attitudes at 2000 ft MSL. The private pilot was recovering the airplane when the instructor told the student "to remove the goggles and that the engine had failed." An attempt to restart the engine was unsuccessful, and the private pilot initiated a forced landing to a road. When it became evident that the private pilot had overshot the landing point, the instructor assumed control of the airplane. During the landing roll the nose gear completely separated from the aircraft, the propeller struck the ground, and the engine firewall sustained structural damage.

I think the instructor showed remarkable restraint in allowing the student to over-shoot the emergency approach before assuming control of the airplane—did things look good to both pilots, until the plane was low enough for their predicament to be obviously visible? In this case, there was plenty of light, but the sun was low in the horizon again. As the plane descended to the ground, the conditions would noticeably darken, as they flew into the shadow of the horizon. Did they have difficulty selecting a landing area, as well as judging the approach?

Chino, California

On October 19, 1997, at 1925 hours PDT, a Beech D17S, operated by the pilot, experienced a total loss of engine power approaching runway 26R at the Chino Airport. The pilot made a forced landing in a plowed field and collided with dirt berms about 1/2 mi short of the runway. The airplane was substantially damaged, and the airline transport-certificated pilot and passenger received minor injuries. VMC prevailed during the personal flight, which originated from Carlsbad, California, at an undetermined time. According to the FAA, during its on-scene examination of the 1943 model "Stagger Wing" airplane, no fuel was found in the two upper wing tanks, the center fuselage tank, or in the lower wing's right tank. Approximately 10 gallons of fuel was observed in the lower wing's left tank.

I find that while driving home at night, I just want to get there, putting off scenery, amusement and service stations for another day. I'm tired and I want to go home. Is a pilot flying at night less likely to land and refuel, for similar reasons?

Emporia, Kansas

On October 23, 1997, at 2017 CDT, a Mooney M20K, operated by a private pilot, was destroyed when, while maneuvering, the airplane impacted the terrain. Instrument conditions prevailed at the time of the accident. An IFR flight plan was on file. The pilot and two passengers on board the airplane were fatally injured. The flight originated at Farmington, New Mexico, and was en route to Emporia, Kansas.

"Destroyed while maneuvering" suggests that the accident occurred when the airplane was low to the ground and probably visual. My guess is that it hit the ground while circling visually at the end of an instrument approach—a difficult prospect when visibility is good. At least while IFR, they probably didn't see it coming.

Fernley, Nevada

On October 24, 1997, at 0250 PDT, a Cessna 170B made a hard landing at Tiger Field in Fernley. The aircraft sustained substantial damage; however, the pilot, the sole occupant, was not injured. The aircraft was being operated as a personal flight when the accident occurred. The flight originated in North Las Vegas, Nevada, at 2330 on October 23. Visual conditions prevailed at the accident site, and no flight plan was filed. The aircraft was making an approach to an unlighted runway.

No lights. Little chance. Refer to the story at the beginning of Chap. 7.

Hazen, Arkansas

On October 24, 1997, at 1913 CDT, a Piper turbo arrow was destroyed during an uncontrolled descent near Hazen, Arkansas. The non-instrument-rated private pilot and the passenger received fatal injuries. VMC conditions existed at 1852 when the flight departed Little Rock, Arkansas. A VFR flight plan was filed for the personal cross-country flight to Murfreesboro, Tennessee. During personal interviews conducted by the investigator in charge, local authorities, witnesses, and the operator reported that the airplane departed Houston Gulf Airport, League City, Texas, at noon, on October 24. The pilot, sole occupant of the airplane, flew to Longview, Texas, where the aircraft was topped with 19 gallons of fuel and the passenger boarded the aircraft. An en route refueling stop was made at Little Rock, where the airplane was topped with 19 gallons of fuel. One witness reported hearing an airplane "motor cut in and out and then heard a loud crash." Another witness observed the airplane flying "about 200 [to] 30 ft elevation going very slowly in an easterly direction. The [air]plane appeared to make a turn to the north. The [air]plane appeared to be going straight down." Witnesses reported the accident, and local authorities initiated a ground and airborne search; however, night, terrain and weather conditions hampered the search. The airplane was located at 1314 the following day on the Wattensaw Wildlife Management Area.

If you crash at night, you'd best be prepared to manage on your own until conditions are light enough for the search party to find you—and try to crash gently, if you must.

Frankfort, Illinois

On October 30, 1997, at 0340 CST, an Aerospatiale AS350B was destroyed on impact with the terrain and a postcrash fire, at the Frankfort Airport. Both pilot occupants sustained fatal injuries. The public use flight was conducted by the Illinois State Police with the mission characterized as an "orientation" flight. The persons on board the helicopter were a civilian contract pilot holding a commercial helicopter and instructor's certificate and an Illinois State Police Trooper who was the holder of a private helicopter certificate. Witnesses indicated that the helicopter was maneuvering in the vicinity of the airport for about 40 minutes prior to the accident. The flight departed Chicago Midway Airport at 0244 in VMC. No flight plan was on file.

SYNOPSES FROM ACCIDENT DATABASE WITH PROBABLE CAUSES IDENTIFIED

That's it for October of 1997—just one month's worth of night mishaps, and fairly typical ones, at that. Most all the preceding synopses are preliminary in nature—so far, the officials have just gathered the facts and questioned witnesses. After investigation, sometimes a cause for the accident can be pinpointed. In some cases, it might be the night conditions, but more often than not, the cause will be something else, and the night conditions will be listed only as a contributing factor. In truth, where the accidents are fatal, there is only so much that a smashed machine can tell, beyond the fact that it hit the ground pretty hard— who knows exactly what went wrong?

The following accident synopses, selected from the accident database, have night-time conditions, or related difficulties listed as a probable cause. Perhaps we can learn from them.

January 18, 1987—Lake Tahoe, California

After departing at night from South Lake Tahoe, California, the pilot contacted ARTCC and requested radar flight following. A short time later, he declared an emergency and said he was reversing course with a loss of engine power. Radar and radio contact with the aircraft were lost as it was returning over a mountain about 10 mi west of the departure airport. Due to its remote location, the wreckage was not examined until it was recovered on 6/4/87. No preimpact part failure or malfunction was found during the investigation. The fuel selector valve was found in the off position.

This crash illustrates at least three key points:

1. The engine *could* quit.
2. Terrain unusable for landing is best avoided.
3. It might be a *long time* before any help arrives.

January 13, 1987—Sacramento, California

The winds at the airport had been reported as strong and gusting. After his passengers arrived, the pilot requested and received taxi instructions to the active runway which was approximately 3700 ft from his parking space. About 2 1/2 minutes after receiving his taxi clearance, the pilot requested and received takeoff instructions. Ground witnesses stated that after the aircraft took off, it climbed to approximately 100 to 200 ft, rolled left, entered a steep dive, and crashed. About 5 seconds later, a postcrash fire erupted. No preimpact part failure or malfunction was found. The gust lock pin was found near the control column. The pin was bent, and the gust lock pin hole in the control column was elongated as though it had been forced rearward with the pin installed. The light condition was dark night.

It appears as though the pilot taxied with the gust lock installed, because of the strong winds, and forgot to remove the pin prior to takeoff. Evidence suggests that the pilot worked mightily with the locked controls—or that the pin and hole were bent in the crash. Since the night cockpit environment is purposely dark to facilitate night vision and the gust lock itself is not lit nor located in a lighted area, is it possible that the gust lock would have been easier to overlook? This one is easily prevented by a simple control function "free and correct" check prior to take-off.

January 16, 1987

While on a Day-VFR flight from Ontario, California, to Las Vegas, Nevada, the aircraft collided with mountainous terrain near the top of a 6500 ft peak. Radar data showed that approximately 20 minutes before the accident, the aircraft began a gradual descent from about 10,000 ft. No preimpact part failure or malfunction of the aircraft, engine, or autopilot was found. The pilot had a rest period on the previous day, but the investigation did not determine whether he had obtained any sleep before a 1700 PST flight. After the 1700 PST flight, he voluntarily flew on a late night flight with another company. That flight was delayed in returning and did not land at Ontario until about 0500 PST the next morning. Approximately 41 minutes later, the pilot took off on the accident flight.

This pilot was tired. For a similar experience, refer to the story that begins Chap. 4.

March 2, 1993—Oakley, Utah

The flight took off at 0500 hours in dark visual meteorological conditions. Radar data indicate that the flight was heading in an easterly direction, toward its destination, and had climbed to and leveled off at 12,500 ft. Three minutes prior to the last identified radar target, the pilot obtained a partial weather briefing, for the destination airports, from the flight service station. The wreckage was located in an area of high mountainous terrain. Evidence indicated that the flight collided with a ridge-line at the 12,400 ft level. The main wreckage was located on the east side of the ridge at 11,400 ft. During the postcrash investigation, there was no evidence of a mechanical failure or malfunction.

The pilot in the right seat of this accident worked as a flight instructor in Salt Lake City. The accident flight was his first day on the job as a pilot with a charter company—the Cessna 402B was full of bank checks and cargo. He and a senior company check pilot departed in darkness, flying VFR, even though the company maintained a "canned" IFR flight plan on file; the IFR plan was never activated. It appears as though they pointed directly at their destination via LORAN, and figured that 12,500 ft would be high enough. Unfortunately, the light conditions of the early morning rendered the high terrain invisible, and their altitude was only close. What a surprise it must have been to suddenly bounce off an unseen mountain. The airplane actually hit the mountain on the west side and bounced over the ridge, coming to rest on its east face. At dawn, a few hours later, everything became perfectly clear, but sadly, the pilots were too dead to see it.

July 4, 1997

At 2100 EDT, a Cessna 170B, piloted by a commercial pilot, was destroyed during a collision with an electrical power transmission line, the ground, and subsequent fire. VMC prevailed at the time of the accident. The flight was not operating on a flight plan. The pilot and passenger were fatally injured. The flight departed Interlochen, Michigan, exact time unknown. The terrain in the area of the accident was hilly. Shortly before the accident, witnesses reported the airplane was flying at tree top level and performing maneuvers similar to an aerial applicator airplane. None of the witnesses observed the airplane's collision with the electrical power transmission line.

Cropdusters use *big* lights to help them see in the dark—suggesting that a typical landing light is not adequate for low altitude maneuvering.

June 29, 1983—Kahului, Hawaii

During arrival, a descent was begun from 8000 to 7000 ft MSL at 0151 HST. At O155 HST, the air crew was cleared for a visual approach from 7000 ft MSL, and a rapid descent was started with 20 degrees of flaps and approximately 25 inches manifold pressure. The aircraft was maneuvered to intercept the localizer and glide slope as a reference. After intercepting the glide slope at about 3000 ft, the pilot in command called for 2250 rpm and gear extension. While descending through approximately 2000 ft, he noticed the aircraft slowing down and descending below the glide slope, so he called for a power increase to 27 inches manifold pressure. At about that time, the air crew noted that the engines were not responding and had lost power. Subsequently, a forced landing was made in a sugar cane field with the gear in a transient position. The fuel selectors were found positioned to fuel tanks containing fuel, but the positioning during the descent was not verified. The temperature and dew point were 72° and 65°F. This would have been barely within the envelope for carburetor ice on icing probability charts.

Because of cooling temperatures at night, there is a greater chance of carburetor ice. Waiting until the ground is in view, during a night emergency landing, may prohibit a successful landing gear extension. If the terrain below is unseen, it is probably best to configure for landing early, and slow the aircraft to a minimum of speed and descent rate.

January 17, 1987

After leaving clouds about 4500 ft MSL, the pilot in command made a visual approach to runway 03 at Lynchburg municipal airport, Lynchburg, Virginia. On short final the copilot stated that he noticed what seemed to be a gradual increase in pitch attitude while the aircraft was still too high off the runway. He mentioned this to the pilot in command; however, the aircraft descended rapidly and contacted the runway, collapsing the left main gear. The aircraft slid a short distance on the runway. Light conditions were dark night.

This pilot flared too high and stalled, a possible victim of an illusion created by runway lights at night—see Chap. 7.

January 20, 1987

The pilot was making a night VFR landing approach to runway 22 at Taos, New Mexico, when the aircraft struck some high-voltage power transmission lines running perpendicular to the runway. The pilot was able to maintain control of the aircraft and landed it successfully. The aircraft was substantially damaged; however, the pilot was not injured.

VASI lights are intended to prevent this sort of accident.

January 18, 1987—Bakersfield, California

A Cessna 152 collided with a dirt bank located approximately 140 ft to the left of the runway edge while on a night approach. The runway safety area at the airport extends 60 ft from the runway centerline. The pilot in command indicated that he was unable to activate the low-intensity runway lights after repeated attempts and that the airport's rotating beacon was inoperative. The pilot in command also stated that the private pilot passenger was manipulating the controls at the time, and he aligned the aircraft on what was thought to be the runway. The pilot-controlled airport lighting system was operated 10 hours prior to the accident with no noted deficiencies. Subsequent testing also revealed no deficiencies. The aircraft radios were not tested because they were destroyed in the postcrash fire. The airport's rotating beacon was reported out of service and was published in the notams. Another airport, located 6 nautical mi north, is equipped with HIRL and MIRL on either of its two runways which operate continually.

Most pilots need to see the runway in order to land the plane. A similar situation is described in the beginning of Chap. 7.

June 19, 1984—Barber's Point, Hawaii

The pilot stated he got off course during the flight because of a faulty ADF. Radar contact was established, and the aircraft was being escorted into Honolulu when fuel exhaustion occurred. The aircraft was ditched approximately 10 mi southwest of Barber's Point Naval Air Station, and the pilot was rescued by the Coast Guard. Light conditions were dark night, and the water was rough.

Lucky pilot. Over water at night makes for instrument conditions—as does flight over large unpopulated areas (no lights). Without instruments, there is a strong possibility of getting lost. More contingency fuel would be nice.

CONCLUSION

There you have it, a short selection of night accidents that offers a nice, broad spectrum concept of ways to go about crashing in the dark. Within these short histories, you may have noticed some similarities. Landing accidents, for example, seem to follow a similar circumstance—the pilot has difficulty judging the approach or interpreting lights. There are other consistencies in the accidents where perfectly good airplanes are flown into unseen terrain.

As I said before, you may draw your own conclusions and, one hopes, avoid making the same mistakes.

About the Author

Lewis Bjork is a pilot for Skywest Airlines. A specialist in aerobatics with thousands of instruction hours, he has flown more than 130 types of airplanes. As an experimental aircraft builder and test pilot, Bjork refined the handling characteristics of a microlight twin and a pressurized Lancair IV. He is the author of *Piloting for Maximum Performance* (McGraw-Hill).

Index

INDEX